BACKYARD
CASH CROPS

The Sourcebook for Growing
and Selling Over 200 High-Value
Specialty Crops

BACKYARD CASH CROPS
The Sourcebook for Growing
and Selling Over 200 High-Value
Specialty Crops

Second Edition

COPYRIGHT © 1992

HOMESTEAD DESIGN, INC.

**PUBLISHED BY: HOMESTEAD DESIGN, INC.
P.O. BOX 1058, BELLINGHAM, WA 98227**

PRINTED IN THE U.S.A.
A special thanks to John Adams, Sheridan Bell, Debbie Emery, and Jean Wallin

ISBN 0-933239-32-7

Wallin, Craig, 1944-
    Backyard cash crops: the sourcebook for growing and selling over 200 high-value specialty crops / [Craig Wallin]. –2nd ed. p. cm.
    Includes bibliographies and index.
    ISBN0-933239-32-7 (pbk.)
    1. Horticultural crops. 2. Horticultural crops–Marketing. 3. Horticulture–Directories.
    4. Horticultural crops–United States. 5. Horticultural crops–United States–Marketing.
    6. Horticulture–United States–Directories. I. Title. II. Title: Back yard cash crops.
    SB318.W35
    635-dc19
                                                                                89-1778
                                                                                CIP

TABLE OF CONTENTS

TABLE OF CONTENTS

TABLE OF CONTENTS

finger·prince

GROWING FOR PROFIT

"Ere long the most valuable of all arts will be the art of deriving a comfortable living from the smallest area of soil. No community whose every member possesses this art can ever be the victim of oppression in any form." . . . . *Abraham Lincoln*

There is nothing like the satisfaction of discovering a way to make money that's enjoyable, productive and honest. In the pages that follow, you'll discover dozens of ways to "grow" money with back-yard cash crops. You will find many advantages to being a small grower:

- You can start with little capital.
- Many crops are recession-proof.
- You can start at home.
- You can run your business in your spare-time.

Whether you're just looking for a way to earn extra money, or want to become a full-time grower, you can find a crop to fit your needs. You do not need a large acreage to grow cash crops. An average backyard can provide a nice supplemental income; and five acres or more may make you wealthy. In fact, many growers have no land, just a basement or garage! (Read about sprouts and mushrooms in Chapter 10).

Whenever possible, figures are given for yield and profit potential for a one-acre planting. Yields per acre and crop prices are based on data supplied by the U.S. Department of Agriculture, grower's trade publications, and individual growers. Cost of outside labor for soil preparation, weeding, and harvesting are not included, as most small-scale growers and their families will prefer to do this work themselves.

Because it would be impossible to provide detailed growing and harvesting information for all the crops listed in less than two thousand pages, a resource list of publications and organizations is included with every chapter to allow you to get current in-depth information on the crops you choose.

Before you begin, be sure that the crops you plan to grow are suitable for your local climate and soil. Talk to other growers in your area and your county extension agent. Local experts can tell you which varieties do best in your area.

FLOWER BULBS

The average bulb grower in Holland sells $140,000 worth of bulbs every year, much of that to cut flower growers who use the bulbs for growing flowers, then toss them out and buy new ones the following year. The Dutch bulb growers are convinced that there is great potential in the U.S. Sales are a modest $20 per person here, much lower than Europe.

Flower bulbs are one of the best crops for the specialty flower growers. Besides being easy to grow, most bulbs multiply rapidly with proper care. "Bulb" is a general term used to describe both true bulbs such as daffodils and tulips, and other underground food storehouses called corms (crocus), rhizomes (iris), and tubers.

For the commercial grower, bulbs that are forced to flower out of season are especially profitable. "Forcing" simulates natural conditions to cause bulbs to bloom months before the normal cycle. The secret of forcing is to plant early enough to allow the bulb time to develop a sturdy root system.

Popular flowers for forcing include: "Paperwhite" daffodil for Christmas. "February Gold" daffodil for January. "Olaf" tulip for Valentine's Day. "Peerless Pink" tulip for Mother's Day. "Anne Marie" Hyacinth for Valentine's Day. "Grote Gele" Dutch crocus for January.

CANNA

These popular large perennials, with their lush tropical foliage and gladiolus-like flowers make borders and planting beds come alive with their vivid colors. Cannas also bloom all thru the summer until the first frost.

Cannas prefer a fertile, well-drained loam. If your soil is poor, it can be improved by adding dried cow manure and bone meal or superphosphate. Plant about two inches deep in a sunny location, with a spacing of 12" to 24" between plants.

Harvest after the first frost, when the leaves die back. The Canna spreads by underground runners called rhizomes. Lift and divide the rhizomes, discarding old growth. Store over the winter in peat moss or sand.

CROCUS

The crocus is the best-known springtime flowering bulb. Hardy in almost all areas, most bloom in earliest spring, but some species will bloom in the fall. The hybrids, also called Dutch Crocus, are the most vigorous and popular for forcing into early bloom in pots. The non-hybrids bloom earlier and have unusual coloration.

Crocus prefer sun or light shade, and a light porous soil. Set the corms 2" to 3" deep and 3" apart. The crocus tends to multiply almost as fast as rabbits, but a commercial grower can accelerate the process even more by root division or by inducing lateral buds. See "Propagation" in the recommended reading section of this chapter.

DAFFODILS

This hardy perennial originated in Europe and has become one of the most popular bulbs, partly because of their virtual immunity to disease and pests. Gophers just hate daffodil bulbs! Daffodils are excellent for naturalizing with vigorous growth, long life, and an abundance of flowers. All daffodils are members of the genus Narcissus, and are usually grouped into twelve divisions according to their shape.

Daffodils are easy to grow. They prefer a well-drained soil and full sun or semi-shade. Bulbs should be planted early in the fall, spaced about eight inches apart and four to six inches deep. They should be mulched in areas with severe winters. Bulbs naturally divide in half each year, so to help the process along, you simply remove one half and plant it elsewhere.

Serious growers should join the American Daffodil Society, at 2302 Byhalia Road, Route #3, Hernando, MC 38632. They publish a quarterly journal for their 1600 grower/members.

GLADIOLUS

A popular cut flower, glads have an extremely wide color range and bloom from spring to fall, depending on the time of planting. The newer varieties of garden gladiolus have spikes and will stand upright without staking.

Glads prefer a rich sandy soil, full sun, and frequent watering. Glads develop up to fifty "cormels", which are miniature corms produced between the new corm and the disintergrating old corm. Collect the cormels when the corm is lifted from the ground before winter. Store them below 40° in a dry frost-free area with air circulating around them. Soak any cormels that have become dry in tapwater for a day before planting.

Cormels will normally take two years to mature. Planting the first bulbs in early spring and then every week will provide blooms through the summer. Plant the corms about four times deeper than their height.

The North American Gladiolus Council has over 2500 members and publishes a quarterly and many helpful publications for growers. Write them C/O Reinhold Vogt, 9338 Manzanita Drive, Sun City, AZ 85373.

IRIS

A large and diverse group of about two hundred species, varying in form, color, growing requirements and methods of propagation. The best known groups are the crested, beardless, and bearded, all three spreading by rhizomes or underground runners. One variety, the roof iris, was traditionally planted in thatched roofs in Japan.

Iris must have a rich, well-drained soil, as it will not tolerate wet feet. Planting depth is critical. The rhizomes should be barely covered with soil. Rhizomes should be planted between July and October, about twelve to eighteen inches apart. The rhizomes grow from the end with leaves. The bearded iris is one of the most prolific plants. The best time to divide the rhizomes is just after flowering. See "Plant Propagation" for specific techniques.

Most commerical growers belong to the American Iris Society, located at 6518 Beachy Ave., Wichita, KS 67206. The society publishes a quarterly, and many books for growers.

LILLIES

One of the most varied garden plants, the lillies are sometimes called "The Glories of the Garden". Their large clusters of brightly colored regal flowers bloom from July through September with a lovely fragrance. About sixty years ago, breeders developed many new hybrids which were healthier, hardier and easier to grow. As a result, it's now possible to grow healthy bulbs in large commercial quantities with a minimum of problems.

Lillies are generally easy to grow. They prefer a deep well-drained soil with ample moisture throughout the year. Planting in filtered light brings out color and makes the blossoms last longer. Plant the bulbs four to eight inches deep, and twelve inches apart.

Growers can join the North American Lily Society, at P.O. Box 476, Waukee, IA 50263.

HYACINTH

This lovely flower originated in the Mediterranean area and is known for it's heavy perfume. The hyacinth is popular for forcing in the winter months. It must have well-drained soil or the bulbs will rot out.

In northern climates, bulbs should be planted in September or October. In milder climates, plant in October thru December. Set the bulbs as deep as their diameter, and six to twelve inches apart. Hyacinths normally multiply too shortly for commercial purposes, so artificial propagation should be used. The two common ways of inducing bulblets are "scooping" and "scoring". Both techniques are illustrated in "Plant Propagation", listed in the recommended reading section.

TULIPS

The tulip was once a holy flower in Turkey and Iran, where it originated. The name comes from the Turkish word for turban. A few centuries later, during the great Dutch "Tulip-Mania", bulb prices spiraled up and up until a single bulb was worth it's weight in gold! Fortunately for us, prices are more realistic now, and everyone can enjoy the tulip rainbow.

Tulips always do best in a rich loam that is perfectly drained. They prefer full sun, but will settle for less.

Planting new bulbs six inches deep and six inches apart will insure both an ample supply of new bulblets and excellent flowering quality for selling cut flowers. Shallow planting leads to undersized bulbs and flowers the next year.

To produce large bulbs for sale in quanitites, you should lift, divide, and store your bulbs each June. Use a garden fork and TLC to avoid injuring the bulbs.

MARKETING FLOWER BULBS

Most small-scale growers have found that quality bulbs at a fair price sell out quickly, Try small newspaper ads and postcard ads on local bulletin boards. Be sure to mention your address and prices! You can sell your cut flowers to local florists and individuals.

• Grade your bulbs by size, and package in clear plastic ventilated bags, (use a paper punch to vent) as pre-packaged bulbs will sell faster. Label each bag with price, color and variety.

• Sell 12 large or 18 medium for the same price.

• Have an instruction sheet for each customer on how to grow great flowers. A satisfied customer will come back next year . . . and the next.

• Offer packages of bulbs that grow and bloom in sequence all spring and summer.

• Give your customers an extra bulb with each dozen. Nothing pleases a customer more than getting something for nothing!

• Sell forced-bulb plants early in the spring through local florists, grocery stores and garden centers.

• Sell your smaller bulbs by the pound to local gardeners who enjoy growing their own bulbs.

- If you decide to specialize in unique varieties, consider mail order. Try a small classified ad in a national or regional magazine.
- In a recent survey of market growers of flower bulbs, the average income per acre was $16,456!

FLOWER BULB SOURCES

DeJaeger Bulbs
188 Asbury Street, So. Hamilton, MA 01982
(free catalog)

Messelaar Bulb Company
Box 269, Ipswich, MA 01938
(free catalog)

John Scheepers, Inc.
63 Wall Street, New York, NY 10005
(free catalog)

White Flower Farm
Litchfield, CT 06759
"The Garden Book" and catalog cost $5 for two issues yearly.

Van Bourgondien
P.O. Box "A", Babylon, NY 11702
(free catalog)

Brecks Dutch Bulbs
P.O. Box 1757, Peoria, IL 61656
(free catalog)

Dutch Gardens, Inc.
P.O. Box 200, Adelphia, NJ 07710
(free catalog)

International Growers Exchange
P.O. Box 52248, Livonia, MI 48152
($5 for catalog)

Iris Test Gardens
1010 Highland Park Drive, College Place, WA 99324
($1 for catalog)

Van Engelen, Inc.
307 Maple Street, Litchfield, CT 06759
(free catalog)

FLOWER BULBS
RECOMMENDED READING

GROWING BULBS . . . E.M. Rix 1978

A commercial grower's handbook, focusing on cultivation in the open garden, in special frames and in greenhouses. Practical advice is given on propagation, pests and diseases, including recommended species. 176 pages.

INTRODUCTION TO FLORICULTURE . . . edited by Roy A. Larsen 1980

This is an excellent text for students and commercial flower growers that covers in great detail the production of commercial floral crops. It provides crop-by-crop coverage of greenhouse and field crops, potted plants and cut flowers. The chapters cover history, taxonomy (plant descriptions) economics, propagation, general culture, disease and insect problems and marketing. 624 pages.

SUCCESS WITH BULBS . . . Park Seed Company, P.O. Box 46, Greenwood, SC 29648

Full of helpful information on growing bulbs. What they look like, how to use them, growing requirements and history.

BULBS – HOW TO SELECT, GROW AND ENJOY . . . George H. Scott, HP Books, 1982

This is THE best book for the beginner and the least expensive in paperback. It provides a complete guide to 250 different bulbs, including propagation, forcing and storage.

ALWAYS DIG BULBS ON A COOL CLOUDY DAY
TO AVOID INJURY FROM THE SUN.

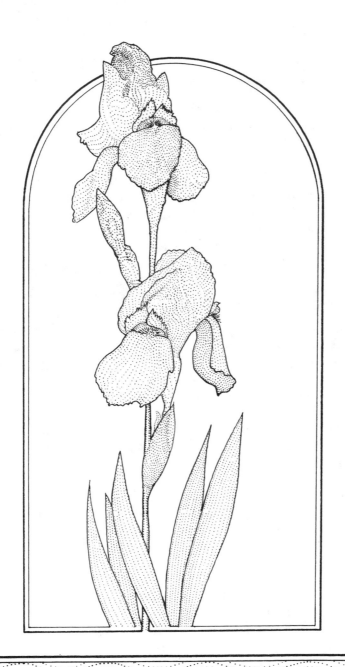

CUT FLOWERS

Growing flowers to sell at the local farmer's market started out as a part-time hobby for Linda Moffit, of Shedd, Oregon. In just three years, her hobby has grown into a full-time summer business with one helper and an acre of flowers in the area's first U-PICK flower patch. Linda grows tiger lillies, larkspur, golden daisies, zinnias, statice, blue salvia, batchelor buttons, glads, dahlias, yarrow, asters, snapdragons and cosmos.

In Holland, where a flower farmer can easily net over $45,000 a year on a quarter acre, there are 16,000 flower growers.

Ted Zwinkels is a good example of a flower farmer. He grows three different types of lilies in a quarter-acre greenhouse, and last year sold $500,000 worth of cut lilies. He times his plantings so that a large crop of premium flowers is ready to harvest when prices are highest — as Americans celebrate Mother's Day.

Because consumer demand can change rapidly, the cutflowers listed in this chapter are those that have enjoyed a reasonably steady demand in years past. However, you should talk to local florists and other growers before making a large investment of time and energy in a particular crop.

Cut flowers sell best in bunches of one dozen. The most effective approach for beginners is to put your cut flowers in a grocery store or florist's shop on a consignment basis. Growers report 90% of their consigned flowers sell within four days, after establishing a working relationship with retailers.

If you are located on a main road, think about a permanent sign advertising your plants, with space on the sign for a special promotion. For example:

SPECIAL TODAY
DAFFODILS $1.49 DOZEN

Remember the peak seasons for cut flowers:
Valentine's Day
Easter
Mother's Day
Memorial Day
Christmas

If your sales volume continues to grow, consider selling wholesale through a commission wholesaler, who usually charges a 20% to 30% commission, or an independent broker who sells to retailers. Both methods allow you to concentrate on growing, and avoid having to hire salaried sales help.

FLOWERS THAT PEOPLE EAT

Prominent chefs have used everything from geraniums to roses to make new and exciting flavors blossom. Edible flowers are **hot**! At top restaurants, it's very trendy to use specially grown flowers to enhance the taste and visual appeal of a fine meal. Now, the folks at Paradise Farms, the pioneer growers of edible flowers, have just published the first book dealing just with edible flowers. It's called "Guide to Cooking with Edible Flowers", and is available direct for just $8, including postage. If you're interested in this lucrative new market for your flowers, order a copy from: Paradise Farms, P.O. Box 436, Summerland, CA 93067.

CARNATIONS

Carnations are one of the best commercial cut flowers, with a strong spicy fragrance and richly colored long-lived blooms. Although they are semi-hardy perennials, they are usually grown as annuals. Because of their fragrance, they are widely grown in Europe for use in cosmetics. They bloom straight throughout the summer in most areas. The taller varieties are usually grown for cutting and the dwarf varieties for plantings.

Carnations prefer a sandy well-drained soil and lots of sun. Start seeds indoors about ten weeks before the last frost, then transplant, using a 6" to 12" spacing.

CHRYSANTHAMUMS

"A chrysanthamum by any other name would be easier to spell."

The chrysanthamum is known to have been cultivated for at least 2500 years. During that time, hybridists have produced a wide range of plant types in an equally wide range of colors. Commercially, it is planted in greater numbers than any other flower. Part of their attraction is late blooming — from late summer through fall — when few other flowers are in bloom. The plants grow as mounds about two feet high and wide, covered with daisy-like flowers.

The giant varieties, commonly called "football" mums because of their size, are in demand by florists for fall corsages. Flowering is controlled by day length, so as summer turns to fall and the days shorten, the flowers bloom.

Chrysanthamums are a slow-growing annual, and should be potted indoors ten weeks before the last average frost. Pinch the growing tips once a month to encourage cushion growth. Mums prefer a loose fertile soil, and should be planted 12" apart in full sun.

GROWING INSECTICIDES

Growers in Oregon are starting to produce pyrethrum plants, which are chrysanthemums with flower heads that look like daisies. The flower head yields a natural insecticide, **pyrethin**, that is extremely safe to warm-blooded animals.

A company called Botanical Resources, based in Corvallis, Oregon, has developed a machine to harvest the flowers at the rate of 5,000 to 10,000 pounds a day, and a new process for extracting the pyrethin from the flowers.

Most of the present supply is grown in Africa, but droughts and lack of irrigation have caused shortages. An agronomist at Oregon State University estimates 10,000 acres will be needed to grow all the flowers needed just for present demand!

The National Chrysanthamum Society, 5012 Kingston Drive, Annandale, VA 22003, with 2800 members, publishes an informative quarterly and a variety of books on chrysanthamums.

ROSES

The rose is one of the most widely grown and appreciated shrubs in America. Beauty is only half it's appeal, with the unique fragrance the other.

During the Middle Ages, roses and other scented flowers were grown within monastery walls and used to decorate the church. In modern times, they have become popular as an expression of love and friendship.

Growing roses is not difficult, but to insure success you must follow four basic rules:
1. Select varieties suited to your local climate.
2. Start with the best plants (#1 grade).
3. Provide basic care: proper planting, water, fertilizer, disease control and pruning.
4. Winter mulch in colder areas.

There are so many variables involved depending on climate and variety that a comprehensive rose book (or two) is a must before beginning. The three types of roses most desirable as cut flowers are GRANDIFLORAS, FLORIBUNDA and POLYANTHA. These types produce abundant clusters of blooms.

SNAPDRAGONS

Like the carnation, snapdragons are perennials, but usually grown as annuals. Native to the Mediterranean, they bloom continually during spring and early summer. Newer hybrids provide new shapes and resistance to heat, allowing extended summer blooms.

Seed indoors approximately eight weeks before the last frost. Transplant 8" to 12" apart in a loose sandy loam with full sun. Pinch the tops of young plants to encourage side branching. If the plants stop blooming during summer's heat, cut back to 6" height for full blooms.

CUT-YOUR-OWN BOUQUETS

Ellen and Shepherd Ogden own the "Cook's Garden", which includes a farm stand, a nursery, and a mail-order seed catalog. When word got around that they had pick-your-own bouquets, a whole new group of customers drove down the dirt road to their homestead, and many returned for vegetables regularly.

Ellen says flowers are a perfect cash crop, because they are easy to grow, produce quickly and supply an income throughout the season. In addition, the startup costs are low, because you only have to buy seeds and supply labor.

Ellen favors annuals, such as calendulas, zinnias, cosmos, snapdragons, bachelor's buttons and asters. She uses plugtrays, a sterile soil mix, and a heating mat to insure quick germination. She says, "Plan your garden for sales. Keep the garden neat and easy for your customers to cut their own flowers and they will come back often. Keep prices reasonable. For every plant in our garden, we try for a yield of $1 per square foot. Advertising 'All you can hold for $3' gets people interested."

RECOMMENDED READING

ANNUALS . . . HOW TO SELECT, GROW, AND ENJOY Derek Fell 1983

A basic guide to over 100 annuals, including over 300 color photos.

ROSES – HOW TO SELECT, GROW AND ENJOY . . . HP Books 1982

This basic paperback is the best rose primer currently available.

COMMERCIAL FLOWER FORCING . . . Laurie, Kiplinger and Nelson 1979

Basic information on plants, soils, pests and marketing of flower crops is covered in this 438 page eighth edition. Considerable attention is devoted to the techniques of cut-flower and potted plant production, including specific propagation and planting information on 25 species of cut flowers, 27 flowering potted plants and 37 bedding plants.

GREENHOUSE FLOWERS & BEDDING PLANTS . . . George S. Williams 1983

This 336 page vocational training text provides the basic information on the production and marketing of flowers and bedding plants.

THE GREENHOUSE GROWER – A CAREER IN FLORICULTURE . . . Kennard Nelson 1977

This text presents the fundamentals of floriculture, ranging from preparing the soil to grading and packing the crop for market. It's written for the person who has little or no previous floriculture experience.

GROWING CHRYSANTHAMUMS . . . Harry Randall 1983

Harry Randall is well-known as an expert grower of chrysanthamums, and in this book he provides a step-by-step guide to all aspects of cultivation. Virtually every type of mum is covered, from grown in the garden to those which flower late in the greenhouse. Abundantly illustrated with color plates and fine drawing, this book is an invaluable guide to both the newcomer and the experienced grower.

THE COMBINED ROSE LIST

Every year Beverly Dobson compiles a list of every species and hybrid rose cultivar available at nurseries throughout North America. Roses are listed by cultivar name and where to get it. In addition, she publishes a bi-monthly newsletter — "Bev Dobson's Rose Letter", just packed with rose news. Both the Annual and the newsletter are a "must have" for serious growers. Beverly R. Dobson, 215 Harriman Rd., Irvington, NY 10533.

CUT FLOWER SEED SOURCES

W. Atlee Burpee
300 Park Avenue, Warminster, PA 18974
(free catalog)

Carter Seeds
475 Mar Vista Dr., Vista, CA 92083
Toll-free (800) 872-7711

Comstock Ferre Company
263 Main Street, Wethersfield, CT 06109
(free catalog)

Farmer Seed & Nursery
P.O. Box 129, Fairbault, MN 55021
(free catalog)

Jackson & Perkins
Box 1028, Medford, OR 97501
(free catalog)

Joseph Harris Company
3670 Buffalo Road, Rochester, NY 14624
(free catalog)

McLaughlin Seeds
Box 550, Mead, WA 99021
(free catalog)

George W. Park Seed Company
Greenwood, SC 29647
(free catalog)

Stokes Seeds, Inc.
Box 548, Buffalo, NY 14240
(specializes in hard to find annuals — free catalog)

Thompson & Morgan
Box 1308, Jackson, NJ 08527
(largest variety of flower seeds in the world — free catalog)

LIVE FLOWERS

FLOWER FARMING BLOOMING!

The fastest growing business in agriculture today is the production of flowers, says Dave Koranski, USDA horticulturist. Koranski says the demand has increased 10% or more each year for many years, partly because of the increasing use of flowers by stores and restaurants.

A good grower with good markets can sell $90,000 worth of plants in one season from a $12,000 greenhouse, he says. Seeds, labor, and other costs total 40% of gross sales.

Paul Brackman, who owns a ten acre nursery that produces only blooming plants at the rate of two million pots per year, has found a wide demand for "instant color" plants, grown in four inch pots and ready to transplant to the garden as finished blooming plants. Paul says, "You have to keep close tabs on demand. You have to be able to sell what you grow. People want that finished appearance when they first plant . . . they just have no patience!"

Paul pre-sells by telephone to independent nurseries and chain stores. Today, he says, "People buy convenience flowers the way they buy TV dinners."

You can take advantage of this demand by focusing on plants that are in demand without competing with the existing volume growers. Find popular plants that take more "T.L.C." than large-scale growers can give.

According to the U.S. Department of Agriculture, standard annuals, the ones customers know best, are usually dependable sellers provided they are in bloom at selling time. These include marigolds, petunias, salvia and zinnias. Begonias and impatients, though more costly to raise, always seem to sell out first.

There is also a market for hardy perennials, particularly if they are brought into bloom during the peak selling season. These include columbine, bleeding heart, delphiniums and bearded iris.

Growers in rural areas should concentrate on growing for resale to garden centers, nurseries, florists and grocery stores. Concentrate on plants the big wholesalers have overlooked because they require special care, but for which there is always a market. Plants such as African Violets are always in demand.

One African Violet grower, Phil Duke, uses the basement of his Chicago area home, and averages $20 per hour for his time. Phil has this to say about commercial growing:
"Most of the small-scale commercial violet growers I know put their mini-farms in their basement. Figure that each square foot of growing space will produce about nine potted plants (in four-inch containers) four times a year."

African Violets are easily propagated from cuttings. Starters are available from several commercial suppliers, including the following.

Alice's Violet Room
Rt.6, Box 233, Waynesville, MO 65583
(S.A.S.E. for free catalog)

Buell's Greenhouses, Inc.
P.O. Box 218/11 Weeks Rd., Eastford, CT 06242
(Catalog 50c)

Fischer Greenhouses
Oak Avenue, Linwood, NJ 08221
(Catalog 50c)

Tinari Greenhouses
P.O. Box 190, Huntington Valley, PA 19006

Zaca Vista Nursery
1190 Alamo Pintado Road, Solvang, CA 93463
(Catalog $1.50)

The African Violet Society of America, with over 28,000 members, publishes "The African Violet" five times a year for amateur and commercial growers. Write them at: P.O. Box 3609, Beaumont, TX 77704.

"HERITAGE" ROSES ARE HOT!

The term "heritage" generally refers to a rose that was in cultivation before 1867, when "LaFrance", the first hybrid tea rose was developed. The hybrids became popular because they were mainly perpetual bloomers and had bolder colors.

Why the renewed interest in these old beauties? There are several reasons. One is the exquisite fragrance of the old roses. Another is their hardiness. The old heritage roses tend to be much more disease-resistant than the modern hybrids.

Many gardeners find the historic lore of the heritage roses appealing. They can raise some of the same varieties that Josephine, wife of Napoleon, grew in her garden at Malmaison, or they can choose the White Rose of York which represented the House of York in the War of Roses.

The appeal of heritage roses is timeless, and the amount of care they require fits the modern gardener's busy schedule.

HERITAGE ROSE SOURCES

HERITAGE ROSES – Miriam Wilkins started this organization for lovers of the old heritage roses and has it grown! Now over 2,000 members receive a quarterly which gives growing hints, historical information, and lists of growers. According to Miriam, "We are a small and informal group working to preserve the classic beauties of the past, and to encourage gardeners NOT to use poisonous chemicals on their property." Miriam Wilkins, 925 Galvin Drive, El Cerrito, CA 94530.

HERITAGE ROSE GARDENS
16831 Mitchell Creek Dr., Fort Bragg, CA 95437
(Catalog $1.00)

HIGH COUNTRY ROSARIUM
1717 Downing St., Denver, CO 80218
(Catalog $1.00)

ROSES OF YESTERDAY & TODAY
802 Brown's Valley Road, Watsonville, CA 95076
(Catalog $3.00)

Rose growers should join the American Rose Society, a group of over 17,000 amateur and commercial growers. Their monthly "American Rose" is available to members. Write to them at P.O. Box 30000, Shreveport, LA 71130.

The Floriculture Department of Cornell University in New York state has done extensive research on profitable potted greenhouse plants, and claims the five most profitable are: poinsettias, lillies, chrysanthamums, geraniums and hydrangeas. You can get the complete report "#AE RES81-21, Economic Analysis of Greenhouse Enterprises" from:
Agricultural Experiment Station
Publications Office
Cornell University
Ithaca, NY 14853
Be sure to get their free list of other helpful publications for growers.

The folks at Nichols Garden Nursery report that there is a growing demand for pansies in the retail and wholesale markets, and that pansies are ideally suited for the small market grower. They have put together an excellent booklet, based on their actual growing experience, that covers all facets of "Profitable Pansy Production" in a concise 10 pages. Write for a copy of their free seed catalog — Nichols Garden Nursery, 1190 North Pacific Highway, Albany, OR 97321.

WILDFLOWERS

The wildflower business may bloom as never before in coming years. Federal legislation recently adopted requires that a percentage of Federal Highway landscaping funds be used to plant wildflowers. Most States already have a native plant program involving highways, following the lead of Texas, which began it's highway beautification with wildflowers in 1927.

A resurgence of interest in the landscaping of the last century, with it's emphasis on native plants, also is noted by Dr. E. "Pat" Carpenter, professor of plant science at the University of Connecticut. Homeowners who want to get the most yard beauty from the least work, he says, are turning to "natural landscaping."

Projects such as these may not only open up new markets for producers of wildflower seeds, but also stimulate greater demand for wildflower plants around homes.

"Wildflowers" is the name for herbaceous flowering plants native to an area. They come in many shapes, sizes and colors and grow everywhere. Although they grow in all states, most are extremely sensitive to the "microclimate" in which they can be grown. Wildflowers have adapted by nature to environmental factors such as rainfall, altitude and exposure to sun or shade. Any attempt to re-establish wildflowers must recognize these factors. So the first thing a grower needs to learn is which flowers are native or suitable to their area.

WILDFLOWER RESOURCES

Growing and Propagating Wild Flowers by Harry Phillips, University of North Carolina Press, P.O. Box 2288, Chapel Hill, NC 27514

Directory To Resources on Wildflower Propagation — National Council of State Garden Clubs, 4401 Magnolia Ave., St. Louis, MO 63110

USDA Wildflower Consultant — Dr. Richard E. Bir, Mountain Horticultural Crops Research Station, 2016 Fanning Bridge Road, Fletcherm NC 28732

Wildflower Research Center — The National Wildflower Research Center in Texas is the heart of research and information about U.S. wildflowers. This nonprofit organization has a membership of more than 8,000. It publishes a quarterly newsletter for members. For membership information, write: NWRC, 2600 FM 973 North, Austin, TX 78725. Please include a LARGE S.A.S.E. with 65c postage.

WILDFLOWER SEED SOURCES

The commercial division of W. Atlee Burpee (300 Park Ave., Warminster, PA 18974) offers bulk seed at wholesale prices for eight wildflower blends. There are mixtures for specific regions, such as their Midwestern blend, the Northeast and Canada blend, and the Southeastern blend. Another mixture is blended just for dried arrangements, and includes seven different varieties. Another mixture was blended to provide especially showy cut flowers, while yet another mixture contains 15 different Perennial wildflowers.

Seed for individual wildflower varieties is available from: Seed Source, Rt. 2, Box 265B, Asheville, NC 28805. They are also a source of bulk perennial flowers seeds. Their current pricelist is $2.

Native Seeds, Inc. offers bulk wildflower seed at wholesale prices for individual varieties as well as regional blends. They will even custom blend for individual states! 14590 Triadelphia Mill Road, Dayton, MD 21036

CULINARY HERBS

Although Europeans use almost five times more cooking herbs than we Americans this is changing fast as cooks here discover the benefits of herbs. Only thirty of the more popular cooking herbs are listed here. The herbal enthusiasts will want to experiment with many more. To give you a glimpse of what other herb growers are doing around the country, let's take a tour.

Marilyn Miller started her herb business, **Oakdale Herb Farm, Rt. 1, Box 48, Blumford, IL 62814,** as a sideline growing for friends, and it rapidly grew into a mail-order catalog business. In 1986, she and her husband Ben built a small commercial greenhouse to grow herb starts for transplanting to the field and to grow and sell herb bedding plants. The cost of the greenhouse and production equipment came to $7,000.

According to Marilyn, "To my complete astonishment, within the first four months of retail sales we had recouped our initial investment and still had plants left for our fields. We also found out that growing herbs in the greenhouse is much easier than growing potted plants, flowers or vegetables. Their growth requirements are not as strict regarding light, moisture and fertilizer."

Marilyn and Ben also have grown five acres of field-grown herbs, using just a Troy-Bilt tiller and cultivator. These herbs are sold in bulk to the culinary, health and natural foods markets and to manufacturers of teas and natural medicinals. The herbs are dried in a 400 square foot loft of their barn. Says Marilyn, "There is nothing to stop you from creating your own niche in the herb world. The market is there, all that is needed is the product."

Quail Mountain Herb Farm, in northern California, grows 15 herb varieties on ten acres for sale to fresh produce market. Their herbs are shipped to produce brokers in major U.S. cities during their main season from October through April. During this seasonal peak, just three people and a rototiller manage the entire ten acres.

"There is money to be made in herbs! It's not uncommon for us to receive orders for 50,000 plants at one time," says Kent

Taylor, at Taylor's Herb Gardens in Vista, CA. With 25 acres of herbs, they ship an average of one million potted herbs per year. Kent also says "The market for fresh-cut herbs for restaurants and gourmet grocers is wide open."

Trout Lake Farm, in Washington state, is one of the largest organic herb farms in North America, with 100 acres of peppermint, catnip for the pet industry, comfrey, lemon thyme, and other herbs. The owner, Lon Johnson, has also started the Organic Herb Growers and Producers Cooperative (Rt. 1, Box 355, Trout Lake, WA) to get the best prices for growers by selling directly to retailers.

In Ohio, grower Jack Schulz has been growing tarragon for resale as one year plants for over 25 years. He grows 90,000 plants on just one acre, from rooted cuttings and root divisions.

Sal Gilbertie, a commercial herb grower in Westport, Conneticut cautions that "People should go into herbs because herbs are what really excite them, not just to make a lot of money. It's got to be fun! I've been in the business thirty years and I still wake up wanting to get to work because I love it!"

Polly Haynes of Rutland, Vermont, started her Meadow Sweet Herb Farm on a shoestring. Says Polly "I put a small ad in the local paper when we opened. We expected a few people to drop in out of curiosity, but before opening time, people were lined up waiting to get in! We were sold out of everything in just one month. The next year we grew twice as much and the herbs only lasted six weeks!"

For those would-be growers who have limited growing space, consider using intensive greenhouse growing, as they do at G.L.I.E. farms in the heart of the Bronx section of New York City. According to the founder, Gary Waldron, the business was started with less than ten dollars and is now a very profitable enterprise, selling 30 varieties of fresh-cut herbs to gourmet restaurants and cooking schools in New York City. Gary had these specific suggestions for other herb growers:

Develop a working relationship with chefs. Suggest new things to them, as they are always looking for a new taste sensation and love to get the jump on other chefs.

Remember you are selling a product AND a service.

Be accurate about the herbs you sell.

The consumer is interested in a year-around supply, not just a seasonal supply.

The more exotic and interesting the product, the more interested the buyers are.

FRESH-CUT HERBS

The demand for fresh-cut culinary herbs is growing every year, as more and more cooks discover the superiority of fresh herbs. A repeat business can easily be established with grocers, caterers and restaurants once you prove yourself a dependable supplier. According to one grower, "You will be selling to a market of people who are accustomed to paying a lot for what they get, but expect the freshest and the best."

As you can see, there are as many ways to make a good living growing herbs as there are growers. Each one finds an approach that suits their skills and facilities. Now lets take a brief look at the basic culinary herbs you can grow for market.

ANISE

Anise is native to the Mediterranean region, and has been cultivated there for thousands of years. In the 5th century B.C. it was recommended by Hippocrates for coughs. The Romans used anise as a digestive aid and to sweeten the breath. They valued it so highly it was used to pay taxes.

The aniseeds have the stongest essence with a flavor similar to fennel seed, as both contain anethol oil. Anise is commonly used today in sweet cakes, soups, beverages such as anisette, medicines and even as an organic dog repellant. Chewing the seeds after a meal sweetens the breath.

Anise, because of it's Mediterranean origins, is a short season crop in North America. Sow seeds directly in May in a light dry soil and thin to 8" apart. Gather the seedheads when the seeds change color. Be sure to put them in a bag so the seeds don't scatter as they fall.

BASIL

Basil is one of the most popular culinary herbs, thanks in part to the current trend towards pasta dishes. There are many varieties from the most popular "sweet basil" to lemon basil, with leaves four inches across. Basil is an annual member of the mint family, producing large crops of richly scented leaves. An acre will produce one to two tons of dried leaf.

Basil prefers a fertile soil with full sun. Thin young plants to six inch spacing, then harvest by regularly pinching off the tops. If you don't the plant will go to seed and stop producing leaves. As basil is a tender plant, you should harvest completely before the first fall frost.

BORAGE

Originating in Syria, borage is now widely cultivated in America. This hardy annual grows quickly and spreads easily. The young leaves are delicious in salads, tasting a bit like cucumbers. Borage leaves are more commonly used in making a tea which sooths the throat. The leaves should be picked after the dew is off the plant, then chopped and dried over low heat. Since the dried leaves tend to absorb moisture, they should be stored in a well-sealed container. An acre of borage can produce a crop of as much as one thousand pounds.

Borage requires a rich loose soil, kept moist. By keeping the area around the parent plant clear of weeds, the borage will self-seed and young plants will sprout around the parent plant. When starting from seed, sow outdoors in a sunny location. Thin the plants to about 18" to 24" apart. Borage is often kept on through the winter as a "window-sill" plant.

CARAWAY

The aromatic caraway seed is a widely-used flavoring for breads, cakes, cheeses and cabbage, and is best known in German and Scandanavian cooking. The largest commercial plantings are found in Russia and Holland. The seeds are known for their digestive properties and stimulate the appetite if chewed before a meal. In earlier times caraway was used in love potions and added to chicken feed to prevent the birds from wandering. Even today, it is given to homing pigeons to insure their return! The roots of the mature plant are often prepared and eaten just like carrots.

Caraway is a biennial, producing seed the second season. Plant the seeds in the fall (best) or early spring and thin the young plants to about 8" apart. The leaves can be picked the first summer for a spicy addition to soups and stews. The seed clusters develop the second summer, and should be harvested as soon as they turn dark. Dry in the sun.

CAMOMILE

Camomile is one of the most popular herbs. The blooming flowers are steeped to make tea. Camomile is also known as the "plant's doctor", as it is said to promote healthy growth in neighboring plants. Common commercial uses include skin creams, shampoos and hair rinses. It is also being used for "No-mow" lawns as they grow in popularity. Roman and German camomile are the most widely known and used of the many species. Both produce an aromatic oil which is concentrated in the flowers.

Roman camomile is a perenial and can be seeded in a light loam, or propagated by dividing the runners in the spring or late summer. German camomile is a hardy annual which can be sown in fall or spring, indoors or out. It does require light to germinate, so do not bury the seed. An acre of camomile will produce approximately 500 pounds of dried flowers.

CATNIP

Catnip is best known for it's wonderful effect on the feline world. Besides being an addiction for most cats, catnip also makes a very good tea herb. Mixed with honey, catnip tea is often used as a cough remedy. Growers should find a ready market for their catnip, especially sewn into catnip mice, since the fresh local catnip will be much more appealing to cats. An acre of catnip will average about one ton of dried herb.

Catnip is an easy-to-grow perennial that prefers a rich soil and straw mulch around the plants. Plants can be started indoors and transplanted to the outdoors, or seeded directly in late spring. Space plants 12" apart in rows 18" apart. Catnip can also be propagated by root division in the early spring. Harvest as soon as the plant is mature, but before the leaves turn yellow. Dry in the shade to avoid losing the volatile oils.

CHERVIL

Chervil is one of the most delicately flavored herbs. It is sometimes called the "gourmet's parsley", being used just like parsley but with a special flavor. In the Middle Ages chervil leaves were used to ease the pain of rheumatism and bruises, and as a cure for hiccups. Chervil is said to be a rejuvenating herb. Used as a culinary herb, fresh is best, so the market grower should extend the season with greenhouse growing.

Chervil is an annual plant and prefers a well-drained soil with filtered sun. Direct seed in late summer for a spring crop. Thin the plants to about 8" apart. Cut off flowers, as they reduce leaf growth.

CHIVES

Chives are an attractive and hardy member of the onion family. The tall green plants grow in clumps, with lavender flowers in summer. Market growers can provide potted chives for their customers who like to keep chives going through the winter on a windowsill.

Start chives indoors in flats or cubes, then transplant to the garden as soon as the ground can be worked. Space about 2" apart in the row. To enlarge your chive bed each year, remove single bulbs from the clump and plant elsewhere.

CORIANDER

One of the most ancient culinary herbs, coriander was recommended by Hipocrates. The first American settlers brought seeds, and the plant has become naturalized in many areas. This hardy annual has a pungent smell when mature. The ripe seeds are used to flavor stews, pickles, marinades and baked goods such as breads, cakes, and cookies. The fresh leaves are also used in curries and chutneys. In Thailand, the roots are crushed and mixed with garlic and used as flavoring. Crop yields are quite variable, ranging from 500 to 2000 pounds per acre.

Corriander prefers light dry soil and full sun. It does not transplant well, so start the seeds outdoors and thin to 8" apart. Harvest the seedpods when the color changes to light brown, and hang to dry as the pods open.

DILL

In addition to it's well known use in pickling, dill adds a unique flavor to stews, vegetables, sauces, and salad dressings. Dill is a hardy annual that closely resembles fennel in appearance. The seeds are used for oil and fragrance, the flowering "umbels" and leaves are used for pickling and seasonings. Dill can be harvested fresh and sold in bunches as "dill weed" for pickling, or dried. Dried yields average 600 pounds per acre.

Dill prefers a sunny location and a fertile, well-drained soil. After the last frost, sow the seeds outdoors, thinning the seedlings to 6" to 8" apart. Dill is quite prolific, and will multiply without much help if left alone for a season. If you plan to sell fresh dill, plant successive plantings every week or two between April and July. The leaves are harvested about two months after planting. The seeds should be harvested when they turn light brown.

FENNEL

Native to the Mediterranean region, fennel was introduced to Europe by the Romans, and to North America by European settlers. This perennial has several varieties with similar flavors, but the common fennel is the most widely grown, as it is hardy in northern climates where it is grown as an annual. Leaves and stems can be picked fresh through the summer. The seeds should be harvested when they change color. The seeds have digestive and stimulant properties as do many of the umbelliferous herbs. In India, fennel is widely used as an after dinner digestive aid. As a culinary herb, fennel is used as seasoning for fat meat, such as pork or oily fish, as well as lamb, chicken and creamy sauces.

Fennel should be sown outdoors in the spring. It prefers a well-drained fertile soil in a sunny location. Thin to about six inches apart. Save enough seed for your next year's crop. Commercial plantings average a harvest of about 1000 pounds of seed per acre.

HYSSOP

This perennial shrub is native to Europe and was brought to North America by the early colonists. Hyssop was mentioned often in the bible as a purifier, and was used to consecrate the alter at Westminister Abbey. As a culinary herb, hyssop is commonly used as a digestive aid with fatty fish, meats and sausages. The tops are used as flavorings for liquers such as chartreuse. The oil extracted from the leaves is used in perfumes such as eau-de-cologne and in medicinal teas for coughs.

Hyssop grows slowly, so seeds should be started indoors if possible to give them a head start. Once your plants are established, you can also propagate by dividing the older plants in the spring or fall. Hyssop prefers a sunny spot and a well-drained soil rich in calcium. Space plants about two feet apart, and harvest and dry the leaves and tops when the plant flowers. An acre of hyssop can produce over a ton of dried leaf.

LEMON BALM

Lemon balm gained it's name from the lemon scent of the crushed leaves. This prolific herb is native to southern Europe, and was introduced to the north by Romans. Lemon balm is a fairly hardy perennial which is easy to grow and tolerant of most soils. The leaves can be used in stuffing, salads and sauces. Tea is made from the fresh or dried leaves, and is one of the more delicious herbal teas. Old herbals mentioned that increased longevity would result from regular use of lemon balm tea.

Lemon balm prefers partial shade and a fertile well-watered location. Start seeds in soil blocks, then transplant when the seedlings are about four inches high, allowing 24" between plants. Once the first planting is established, propagation by cuttings or root division is an easy way to expand the crop.

Root division can be done in the spring, allowing at least three buds to each piece of root. Lemon balm can be harvested two to three times a year under ideal growing conditions, cutting back the entire plant two inches from the ground. Harvest just as the flowers begin to open, and dry quickly in the shade to preserve the color. Commercial plantings average 1200 pounds of dried leaf per acre yearly.

LOVAGE

Lovage is a large perennial plant with celery-like stalks and umbrella shaped seed heads reaching as high as six feet. The leaves are used in soups, stews and salads, while the seeds add flavor to cheese dishes, breads and pastries. Lovage was grown in most Monastery gardens during the Middle Ages for use as a digestive aid, deodorant and antiseptic. With proper care, the plants will live for more than twenty years.

Lovage is easy to grow, and prefers a sunny spot, fertile soil and good drainage. Sow the seeds in late summer, transplanting the seedlings to a permanent location in late fall or early spring. Because the plants are so large, they will need to be spaced three to five feet apart.

Dividing the lovage roots every three or four years in early spring will keep the plant vigorous. Be sure each piece of root has a bud or "eye", much like a potato. Fresh leaves can be picked in season, or the plant can be cut back to 12" high two or three times yearly. Dry in the shade with ample air circulation, then store the leaves in an airtight container. The seedheads should be harvested just as they are turning brown.

MARJORAM

Sweet marjoram is a low-growing herb with a delicate flavor similar to oregano. Of the dozens of varieties available, the sweet marjoram (origanum marjorana) is the more popular variety for culinary use. In most parts of North America it is an annual, to be seeded each year.

Marjoram prefers a sunny well-drained location. Start indoors and transplant after the last frost, spacing the plants 6" apart. Marjoram's shallow root system requires rich humus and freedom from competing weeds to thrive.

Harvest when the flowers first appear, cutting back to one inch above ground. This will stimulate a second, lusher crop, which also should be harvested at flowering. Dry the harvest until the plants can be sifted through a 1/8" mesh screen and discard the woody stems.

MINT

The hardy mint family, including applemint, orange mint, spearmint and peppermint, are vigorous plants and tolerate shady spots and cold weather well. Because of it's vigor, it can be harvested as soon as it gets six inches high, pinching off the tops as the plants grow. The mints are a favorite flavoring for teas, medicines, cosmetics and candies. The large commercial growers harvest peppermint and spearmint for their oil , averaging over 50 pounds of oil per acre. Small growers should focus on their local markets for fresh mints, potted plants and "value-added" mint products such as jellies and herb vinegars. Mint is traditionally served with duck, lamb and young vegetables. One member of the mint family, pennyroyal, is used in sachets to repel moths and added to dog beds to discourage fleas.

Mints prefer fertile soil and ample water. They can be propagated from cuttings or runners. Plant 6" pieces of the root stem 2" deep horizontally in the spring, or purchase plants of the specific variety you want to establish. Mint, like bamboo, needs to be confined because of it's spreading habits.

Space the plants or cuttings about two feet apart to allow the runners to fill in between plants. To harvest, cut the stems off 2" above ground when the flower buds form. Depending on the season, two or three cuttings per season are possible. Strip the leaves from the stem and dry in a well-ventilated, warm and shaded place. Commercial plantings average 1500 pounds of mint per acre.

OREGANO

Oregano is almost as important as tomatoes in Italian cuisine. This popular culinary herb offers an unforgettable taste and aroma. A hardy perennial from the marjoram family (origanum vulgare), oregano is also used as a garnish for beef and lamb stews, gravies and soups. Oregano oil is also used a toothache remedy, applying a few drops to the affected tooth. When your herb garden is well-established, consider packaging an "Italian Blend" of spices for pizza, pasta and spaghetti sauce, including oregano, marjoram, rosemary, basil and sage.

Oregano, with it's Mediterranean origins, prefers a sunny location with well-drained soil. Once the plants are established, you can pick every week or harvest the whole plant when the flowers appear. Dry outside in the sun, rubbing through a 1/8" screen to process for use. Store in an air-tight container.

PARSLEY

Parsley is so well known that it is often considered more a vegetable than a herb. The ancient Greeks wove it into victory crowns for the athletic games and fed it to racehorses to make them run faster. Today, parsley is widely used as a decorative herb, in bouquet garni, and in soups, stews, sauces and stuffings. The leaves are known as a breath sweetener and rich in vitamins and minerals.

Parsley prefers a fertile moist location with ample sun. Although it is a biennial, parsley is usually grown as an annual because the first year's leaves are better tasting. If you start parsley indoors, be sure to use soil blocks or cubes, as it is sensitive to transplanting. When possible, sow seeds outdoors, allowing about six weeks for germination. Thin to a ten inch spacing when the plants are three to four inches high. Market growers should consider putting the transplant thinnings in four inch pots for sale at transplant time.

ROSEMARY

This evergreen shrub is native to the Mediterranean region, where it will reach a height of six feet and live as long as twenty years. Rosemary is one of the most popular culinary herbs, used in soups, stews, sauces and meats. It is also commonly used in sachets, as an insect repellant and as an ingredient in herbal mouthwash. Many herbal hair conditioners contain rosemary, which gives hair body and luster, stimulates hair roots, and helps control dandruff. A formula for basic herbal hair rinse is listed in the section on fragrant herbs.

Rosemary prefers a sunny location and well-drained, slightly alkaline soil. Because of it's Mediterranean origins, it does not tolerate heavy frosts or temperatures below 28°. In most regions of North America it should be grown in a greenhouse or in moveable containers so the plants can be moved inside in the winter.

Rosemary seeds germinate poorly and grow slowly, taking up to three years to produce a usable plant. Most growers use cuttings or layering to propagate the plant. To propagate by cutting, cut the end tips of new growth in August, burying the bottom four inches in sand. Cover with a clothe or hot-cap for two to three months, then transplant into a pot. To propagate by layering, just weight down lower branches to the ground and cover with an inch of soil. Cut and transplant after new roots are well established.

SAGE

Garden sage is a hardy perennial member of the mint family. There are several varieties available, but the most commonly grown is S. Officinalis, with purple flowers. Sage is an important culinary herb, widely used in stuffings, soups, gravies, meats, eggs and sausage. Sage jelly has a uniquely delicious flavor. As a medicinal herb, sage is used to make tooth powders as an antiseptic gargle for sore throats. Used as a hair rinse, a sage infusion can darken greying hair.

Although sage will thrive for years, new plants should be started every year or two, as the stems become woody after the third year. Sage is easy to start from seed, cuttings, or layering. Because some varieties do not breed true from seed, propagate favorite plants from cuttings or layering. Seeds can be started indoors in March for a four week head start on the growing season. Otherwise, plant outdoors in a sunny area with loose soil rich in calcium. Thin plants to a 12" to 18" spacing. The first year, harvest the leaves and uppermost stems in September to insure winter survival. The second season you can plan on two cuttings. Dry the leaves in a shady spot, then sift through a fine screen for seasoning. Commercial plantings can yield up to 2000 pounds per acre.

SAVORY

Members of the mint family, summer savory, (an annual) and winter savory, (a hardy perennial) are similar in their peppery flavor. Both are called bean herbs because they complement the flavor of all types of beans. Summer savory is used with sausages, soups, stuffing and teas.

Savories do best on well-drained soil in full sun. Plant summer savory outside in spring, allowing about a foot between plants. It requires ample moisture and modest amounts of fertilizer.

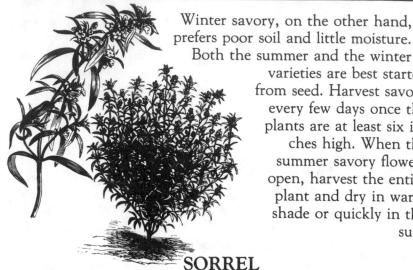

Winter savory, on the other hand, prefers poor soil and little moisture. Both the summer and the winter varieties are best started from seed. Harvest savory every few days once the plants are at least six inches high. When the summer savory flowers open, harvest the entire plant and dry in warm shade or quickly in the sun.

SORREL

French sorrel is the most widely cultivated of the sorrels, with it's large tasty leaves and a tangy flavor. It is high in vitamin C and often used as a spinach substitute. The freshly harvested leaves are also delicious in salads.

Sorrel prefers a sunny spot with dry fertile soil. Because sorrel is a perenial in most areas, propagation can be done with root division or by seeding. Sorrel grows well from seed and can be planted indoors, for a head start on spring, or outdoors in March or April. Space plants six to twelve inches apart. Harvest the leaves continuously through the growing season, as fresh leaves are best.

TARRAGON

Hippocrates, the "Father of Medicine", was prescribing tarragon for a variety of ailments over two thousand years ago. In the Middle Ages, pilgrims put sprigs of tarragon in their shoes for increased endurance on their journeys. Today, this hardy perennial is one of the most popular culinary herbs. It enhances the flavor of other herbs and gives a unique flavor to egg and fish dishes, salads and sauces. Of all the herb vinegars, perhaps none is better known than tarragon vinegar. Tarragon is also an essential ingredient in fine tartar sauce and sauce Bernaise.

The tarragon sold as a herb in the market is French tarragon, which must be grown from cuttings or purchased plants. If you find tarragon seed advertised, it is Russian tarragon (A. Dracunculus) which has almost no flavor. Plant French tarragon in a fairly rich well-drained soil with ample sun. Space plants 12" apart. Harvest in July when the flower leaves turn yellow. Dry the leaves in a warm shady area to avoid browning.

Propagate tarragon from stem cuttings of new growth in the spring and plant in a moist shady place. Cover with a cloche or hotcap. Tarragon can also be propagated by root division. In the spring, divide the root cluster into three clumps and replant. This will also insure vigorous plants.

THYME

Another member of the mint family, thyme has dozens of variations in shape, texture and flavor. Common thyme (T. Vulgaris) is the most widely sold for seasoning. Other varieties are used as ornamentals in rock gardens and as landscaping ground covers. Because thyme keeps it's aroma when dried, it is an excellent winter herb for flavoring bouquet garni, soups and stews. The preservative qualities of thyme also make it useful in sausages, salamis, cheeses and vinegars. This preservative oil, called thymol, also is useful as an antiseptic for mouthwashes as well as for bath oils and soaps. Thyme is often used as a ground cover in orchards to attract bees for pollination.

Thyme prefers full sun and sandy, dry soil with average fertility. Because thyme is a perennial, it can be propagated by seed cuttings, layering or root division. Most commercial growers prefer the ease of starting from seed. Start seeds indoors, as thyme requires a 70° soil temperature for germination. Transplant outdoors in late spring, spacing the plants about ten inches apart.

Harvest the entire plant when the flowers begin to open. Leave the second growth to encourage overwintering and mulch in late fall. Dry the harvest in warm shade. When dry, rub together to separate leaves from stems. Store the dried leaves in an airtight container. Commercial plantings can yield up to 2000 pounds per acre.

SELLING YOUR HERBS

If you're ready to enter the wholesale market with your fresh-cut herbs, you should know about the "Packer's Red Book" (P.O. Box 2939, Shawnee Mission, KS 66201).

The Redbook is a combined directory and credit rating service for the produce industry. The Redbook lists over 12,000 growers, shippers, wholesalers, chain stores and co-op grocers. The semi-annual book includes a weekly update bulletin. Information in the Redbook includes company personnel, net worth of the company and how fast (or slow) they pay their bills. One herb grower wrote a form letter to wholesalers listed in the Redbook and got more orders than she could fill.

Like the Packer's Redbook, the Produce Reporter (315 W. Wesley St., Wheaton, IL 60187) publishes a 1000 page semi-annual "Bluebook" which also includes a weekly update bulletin.

Paula Winchester, who started the successful herb marketing co-op "Herb Gathering" (5742 Kenwood, Kansas City, MO 64110) sells fresh herbs to restaurants. According to Winchester, "When I visit a client, I always feel the fresh herb samples speak for themselves. Herbs are a visual treat to every chef. They start his/her creativity flowing along with their salivary glands. You will need to be aware of culinary trends, so subscribe to magazines such as *Bon Appetit* and *Gourmet,* and trade publications like *National Restaurant Association News.*"

Winchester says growing culinary herbs would appeal to the small intensive farmer who can grow high quality produce and would enjoy the contacts with professional chefs. She stresses that a grower of fresh herbs must live near a large metropolitan area to have easy access to the marketplace.

GREENHOUSE GROWN HERBS

Buyers of culinary herbs, such as grocery stores, delicatessens and restaurants demand a year-round supply of fresh herbs, so greenhouse production is necessary to keep customers satisfied, and money in your bank account! Although field production is more profitable because of lower costs, greenhouse production allows sales that would not be available otherwise.

Herb production in the greenhouse is similar to growing greenhouse ornamentals. To keep heating costs low, you might want to concentrate on cool season herbs that can be grown at 45-55 degrees, such as dill, fennel, parsley, sorrel, rosemary and thyme.

A greenhouse opens up other markets too. Herbs can be sold as bedding plants in the spring. Cut culinary herbs are in demand all winter, and particularly during the holiday season. Many growers report success selling potted herbs such as rosemary as an alternative to poinsettia.

AMERICAN HERBAL PRODUCTS ASSOC.

Growers looking for a market for their crops, or simply ideas on what can be created using herbs should join the American Herbal Products Association. Write for current membership information to: 215 Classic Court, Rohnert Park, CA 94928

TWELVE WAYS TO SELL YOUR HERBS

1. Sell potted annual herbs in spring.

2. Wholesale your potted herb plants to florists, garden centers and supermarkets in the spring.

3. Sell hard-to-find herb seeds thru mail-order ads in gardening magazines.

4. Combine wood indoor window boxes with a "cooks assortment" of culinary herbs in 4" pots.

5. Ornamental culinary herbs in small pots covered with florist's foil are popular sellers at Thanksgiving, Christmas, Valentine's Day, Easter and Mother's Day.

6. Supply your local pet stores with fresh catnip mice.

7. Rent a booth at the county or state fair to sell your herbs or herb products.

8. Retail your fresh-cut herbs at Farmer's markets, craft fairs and through local stores.

9. Retail dried culinary herbs such as savory, rosemary, sage, thyme and tarragon.

10. Talk to local food retailers about setting up a permanent herb rack to merchandise retail packets of dried herbs.

11. Retail herb teas and spice mixes, such as "bouquet garni", herb vinegars, jams and jellies.

12. Organize "Herb Parties" like tupperware parties.

HERBAL PRODUCTS

BASIC SEASONED SALT

1 pound salt
1/4 cup ground black pepper
1/4 cup ground coriander seed
1 tbsp ground bay leaves
1 tbsp ground cloves
1 tbsp dried basil

Mix all ingredients together well. Store in an airtight container, making sure all ingredients are ground fine enough to use in a salt shaker. Vary ingredients to suit your own taste. A blender works well for grinding small quantities of dry herbs. Try adding paprika for a pleasing color.

HERBAL SALT SUBSTITUTES

For those on low or no-sodium diets, consider selling these herbs as substitutes. Summer savory, lovage and celery all have a salty flavor. Dry well and grind to a powder, then store in an airtight container.

ITALIAN SPICE BLEND

Mix equal parts of oregano, marjoram, thyme, rosemary, basil, sage and savory. Vary ingredients to suit your own preference. Store in an airtight container.

HERB MIXTURES

Fines Herbs is a blend of herbs used in cheese and egg dishes, such as omelettes, with white fish and in sauces. Combine 1/2 cup chopped parsley, 1/2 cup chopped chervil, 1/4 cup basil and 1/4 cup thyme. Best if fresh, but can be sold and used dried also. Be sure to store in airtight container.

Bouquet Garni is a well-known blend of herbs used in soups and stocks. There are many variations possible, so do experiment with different combinations. Mix 1/2 cup dried parsley, 1/4 cup dried thyme, 1/4 cup dried marjoram, 2 tbsp ground bay leaf and 1/4 cup dried celery. Cut cheesecloth into four inch squares. Add one level teaspoon to each square and tie into bags. Yield is approximately fifty bags. To sell, heat-seal four to twelve bags in a plastic bag to insure freshness.

HERBAL OILS

Herb oils add a rich flavor to dressings, marinades, and for basting meat and fish. Herbs that are especially good for oils are basil, fennel, marjoram, rosemary, savory, tarragon and thyme. To make herb oils, lightly crush the herb and loosely fill a glass jar. Fill with oil and cover the top with a cloth. Set the jar in a warm (but not hot) place and stir daily. After two weeks, strain the oil, pressing the herbs to extract all the liquid. Pour into clean dry bottles, seal and label.

HERB VINEGARS

Herb vinegars are quite simple to make and add a new dimension of flavoring to dressings, mayonnaise, marinades and flavorings. Suitable herbs include basil, (especially purple basil) chervil, dill, fennel, marjoram, the mints, sage, savory and tarragon.

Vinegar will absorb the flavor of most herbs in two to four weeks. Use a large glass bottle or jar with a non-metal cap to avoid a reaction with the vinegar. Fill the bottle or jar loosely with fresh herbs, then top off with a good quality vinegar. When the vinegar is flavored, strain off the herbs and pour the vinegar into smaller bottles. Add a sprig of the same herb used for flavoring to enhance the appearance, and label each jar.

HERB JELLY

Herb jellies are popular herbal products, and help insure repeat customers. The mints, sages, (especially pineapple sage) tarragon and thyme all make delicious jellies. Here's a basic recipe:

    2 cups herbal infusion
    1/4 cup vinegar
    2 1/2 cups honey
    1/2 bottle fruit pectin

The herb infusion is made by pouring 2 1/2 cups of boiling water over 8 tbsp of fresh herbs. Let it stand for 15 minutes, then strain off the herbs and add vinegar and honey. Bring to a boil, stirring constantly. Add the fruit pectin and boil hard for one minute. Remove from heat, skim, and pour in jelly jars. Add a sprig of the same herb to enhance the appearance and label.

HERB TEA

Once obscure and hard to find, herb teas are now sold in most supermarkets. Because of this increased awareness and appreciation for the unique flavors available in herbal teas, most growers should have no problem finding a ready market for all they care to produce. The three big advantages a "backyard" grower has over the supermarket brands are: freshness, purity and price. Small growers can get their herbal blends to market sooner, guaranteeing a fresher product. They can control growing conditions, insuring herbs that are not contaminated by the chemicals that may be found in more commercial products. Without the overhead of a large corporation, they can offer more attractive prices and quantities to suit their customer's preferences. Some of the more common herb teas include the mints, lemon verbena, camomile, rosemary, lemon thyme, lemon basil, borage, sage and catnip.

$15,000 PER ACRE PLAN

Since culinary herbs sold in stores are usually expensive and not too fresh, you can fill a definite need here. Have a printer make up some heavy cardboard display cards. Put your herbs in heat-sealed plastic bags (1/4 to 1 ounce size) and staple them to display cards. Using this 'bulk pack" approach, customers get a bargain, and you will sell lots of herbs. An average acre should yield enough herbs for 1000 display cards.

SELLING YOUR HERBS WHOLESALE

If you are growing large quantities of herbs and want to explore the possibilities of selling your crop on the wholesale market, or as a contract grower, here's where to find buyers for your crop.

The **International Growers and Marketers Association** serves growers with a newsletter, special seminars and workshops, a directory of growers and buyers, a member's only guide to supplies and products unique to the herb industry and an annual conference. Last year's conference covered such diverse topics as *Advanced Greenhouse Growing, Growing for the Dried Flower Market, Hot New Botanicals, Trends and Pricing in Fresh Cut Herbs.*

For those not able to attend, audio tapes on over 50 topics are available at a modest cost. For information on membership, write to: International Herb Growers and Marketers Association, P.O. Box 281, Silver Springs, PA 17575

The **Produce Marketing Association,** at 1500 Casho Mill Road, Newark, DE 19714-6036 302-738-7100, can provide herb market bibliographies from their computer database to nonmembers for a small fee.

The Agricultural Marketing Service of the U.S.D.A. provides daily price and market reports through it's Market News branch. Reports include fresh-cut herb prices and shipment sizes in New York, Boston, Miami, San Francisco & Los Angeles, the centers of greatest herb use. Write to Market News, A.M.S., Room 2503-S, U.S.D.A., Washington, D.C. 20250

Another office of the U.S.D.A., the Foreign Agricultural Service, supplies information on herbs, spices and processed herbal products in the newsletter "Spices and Essential Oils." $5 brings you a subscription. Write to: F.A.S. Information Division, Room 4644-S, U.S.D.A, Washington, D.C. 20250-1000

The **Chemical Marketing Reporter** (Schnell Publishing, 100 Church St., New York, NY 10007) publishes the latest information on processed herb prices and dealers. Their annual, the *Oil, Paint and Drug Chemical Buyer's Directory*, lists all current dealers.

The quarterly, **Herb, Spice, and Medicinal Plant Digest** ($6 year from Dept. of Plant and Soil Sciences, Stockbridge Hall, University of Mass., Amherst, MA 01003) is filled with solid information on growing and marketing herbs.

RECOMMENDED READING

Serious growers will want to join the American Herb Association, at P.O. Box 353, Rescue, CA 95672. Write for current membership information.

The essential publication for beginning market growers is an excellent newsletter called: "The Business of Herbs", Northwind Farm, Rt. 2, Box 246, Shevlin, MN 56676

SUCCESS WITH HERBS George Park Seed Co., 1982 Greenwood, S.C. 29647

This is a complete guide for the home gardener and commercial grower. Written by a mother and daughter team with an extensive herbal background, the book includes many color photos of herb seedling, a valuable aid to novice growers.

HERB SOCIETY OF AMERICA Two Independence Court,

Concord, MA 01742, offers a variety of publications, including "Primer of Herb Growing."

BROOKLYN BOTANIC GARDENS, *at 1000 Washington Ave., Brooklyn, NY 11225, has an excellent "Handbook on Culinary Herbs", with basic information on selecting, growing, propagation, harvesting and drying herbs.*

HERBAL BOUNTY, *by Steven Foster, contains extensive coverage of herb cultivation, including propagation, culture, culinary uses and medicinal uses. Steven also publishes the* **Herb Business Bulletin,** *a quarterly which covers growing, marketing and research tips. Write for current subscription information to: P.O. Box 32, Berryville, AR 72616*

RODALE HERB BOOK . . . *This comprehensive book, put together with the usual Rodale thoroughness by a group of well-known herbal growers, covers it all. From the teachings of Hippocrates to companion planting, this would be my choice if I could afford just one herb book.*

HERB GROWING AT IT'S BEST . . . *Written by a commercial grower, this book organizes herbs by propagation methods and gives clear instructions for growing both common and uncommon herbs. By Sal Gilbertie.*

GROWING AND USING HERBS SUCCESSFULLY . . . *This practical book was written by a Canadian commercial herb grower, Betty E.M. Jacobs, and covers small-scale commercial herb production techniques. Garden Way*

CULINARY & MEDICINAL HERBS *Published by the British Ministry of Agriculture, this book is aimed at the commercial herb grower. Covers in great detail the production of field grown herbs for the fresh, dry and essential oil markets. Includes sections on propagating, weed and disease control, yields, sowing and planting distances. This is the most comprehensive book of it's kind currently available. Order from: Richters, Goodwood, Ontario, Canada LOC 1AO*

CULINARY HERB SEED SOURCES

Abundant Life Seed Foundation, P.O. Box 722, Port Townsend, WA 98368

This non-profit foundation offers a wide variety of seeds and books and publishes a newsletter for members. They offer wholesale quantities and prices to larger growers. Send $1 for their current catalog.

Catnip Acres Farm, 67 Christian St., Oxford, CT 06483

The display garden here has over 400 herbs, and most are available by mail, including 75 varieties of scented geraniums! The informative catalog costs just $1.

Fox Hill Farms, P.O. Box 9, Parma, MI 49269 *If you can't attend the annual "Basil Festival" at Fox Hill Farms, at least send a dollar for their comprehensive catalog which includes 13 kinds of basil and over 350 herbs.*

H.G. Hastings Co., 434 Marietta St. N.W., P.O. Box 4274, Atlanta, GA 30302 *Specialists in varieties adapted to the warmer southern climate. Free catalog.*

J.L. Hudson, Seedsmen, P.O. Box 1058, Redwood City, CA 94064 *Specializes in rare and imported species of culinary and medicinal herbs. Send one dollar for current catalog.*

Johnny's Selected Seeds, P.O. Box 2580, Albion, ME 04910 *Specialties are organically grown seeds for northern climates. All herb seed is available in commercial (bulk) quantities, and the catalog is free.*

Nichols Garden Nursery, 1190 Pacific Highway N., Albany, OR 97321 *Sells a diversity of products, including herb seed, garlic and shallot bulbs, herb plants, culinary herbs and spices, seasoning blends, fragrance oils and a good selection of herb books. Their catalog is free.*

Peace Seeds, 2385 S.E. Thompson Rd., Corvallis, OR 97333 *One of the widest selections available anywhere, including over 300 varieties of mint and 30 different sages. All are organically grown, and the seed list is free.*

Redwood City Seed Co., P.O. Box 361, Redwood City, CA 94064 *Many herb seeds & rootstocks are available here, including an interesting collection of dye plants. Wholesale quantities are available. Current catalog is $1.*

Richters, Goodwood, Ontario, Canada LOC 1A0 ($3 catalog)

San Francisco Herb Co., 250 14th St., San Francisco, CA 94103

Sandy Mush Herb Nursery, Rt. 2, Surrett Cove Rd., Leicester, NC 28748 *offers an enormous (over 700 varieties) of herbs. The catalog — almost a book — is $4.*

Taylor's Herb Gardens, 1535 Lone Oak Road, Vista, CA 92083 *One of the largest wholesale growers of herb plants. They also sell seeds for over 70 varieties of hard-to-find herbs. The current catalog is free.*

Thompson and Morgan, P.O. Box 1308, Jackson, NJ 08527 *Every serious grower should have a copy of this 200 plus page catalog featuring over 4000 varieties of flowers, herbs and vegetables.*

Space limitations prevent listing the many smaller specialty suppliers to the herb trade. For a comprehensive listing of suppliers, get a copy of the "Herb Gardener's Resource Guide", by Paula Oliver, available from: Northwind Farm, Rt. 2, Box 246, Shevlin, MN 56676

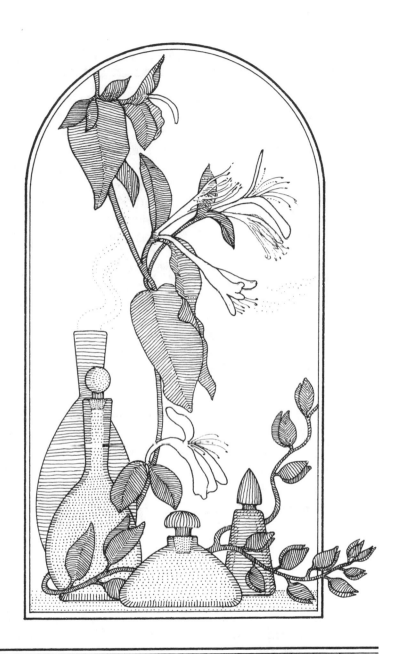

FRAGRANT HERBS

For centuries, the fragrant herbs have been extensively used for their scent. The Greeks wove garlands of marjoram to invigorate guests, while the Romans placed bowls of mint in bedrooms to refresh the senses. During the Middle Ages, a bunch of lemon balm was sniffed during a long sermon to avoid drowsiness! Potpourri and sachets were used to mask unpleasant smells and to prevent the spread of disease. Modern researchers have since found that many of the scented herbs are strong germicides.

Fragrant herbs that are suitable for growing in North America include bergemont, calamus, catnip, camomile, lavender, lemon balm, lemon thyme, lemon verbena, apple mint, orange mint, peppermint, pineapple mint, spearmint, pennyroyal, rose, rosemary, rue, sage, scented geraniums, southernwood, tansey and wormwood. Those who become fascinated by fragrant herbs will find dozens of lesser-known varieties to add to this abbreviated list.

An almost endless variety of herbal products can be crafted using the fragrant herbs. At least one basic recipe has been included in each category to get you started. For more recipes and more complete information, refer to the recommended reading section.

POTPOURRIS AND SACHETS

Natural fragrances are enjoying renewed popularity, with potpourris selling well. Ingredients are similar for both, but a sachet typically uses ground ingredients (an electric blender works well) while potpourris use coursely crushed ingredients.

Most potpourris and sachets have three basic parts: the main scent, the secondary scent and a fixative. Here is a basic recipe, using the three parts.

2 cups dried main scent flowers, such as rose petals or lavender blossoms.
1½ c. dried secondary scent flowers, as violets or carnations.
½ cup dried herbs, such as mint.
¼ cup crushed spices, such as cinnamon, cloves or nutmeg.
¼ cup dried citrus peel (orange of lemon).

½ oz. dried crushed orris root (fixative).
Optional: a few drops of essential oil. Use an eyedropper so you don't add to much oil.

Mix all the ingredients together and store in a large plastic bag in a dark cool place for about four weeks. Stir or shake once a week.

Don't be afraid to experiment by varying quantities and trying other ingredients from your garden. Use bright colorful flowers to add color to the blend.

Dry plant material as quickly as possible to preserve color and fragrance. Use a regular oven, set on warm, with the door cracked slightly to allow moisture to exit. A quicker drying method is to use your microwave oven. Place the plant material between layers of paper towel and microwave on high for a minute or two. When dry, plant material should be crispy dry.

Here are sources for potpourri and sachet ingredients:
Nichols Garden Nursery
1190 N. Pacific Hwy., Albany, OR 97321 (free catalog)

Reminescent Herb Farm
1344 Boone Aire Rd., Florence, KY 41042 (free catalog)

San Francisco Herb Co.
250 14th St., San Francisco, CA 94103
(free catalog) 800-227-4530

MOTH-REPELLANT SACHET
    2 cups dried mint
    2 cups dried rue
    1 cup dried southernwood
    1 cup dried rosemary
    1/4 cup powdered clove
    1/2 oz. orris root (fixative)

HERB GARDEN
    2 cups dried thyme
    1 cup dried rosemary
    1/2 cup lavender

    1 cup dried mint
    1/4 cup dried tansey
    1/4 cup powdered clove
    1/2 oz. orris root (fixative)

FLORAL BLEND

    1 cup dried lavender
    1 cup dried lilac petals
    1 cup dried rose petals
    2 cups dried rose leaves
    1/4 cup cinnamon bark
    1/4 cup powdered clove
    1/2 oz. orris root (fixative)

HERBAL SOAPS

Making scented soaps is quite simple. You can start from scratch with an old-fashioned lye soap, (basic recipes can be found in the Rodale Herb Book). A popular herbal product, the herbal "wash ball", used mainly for washing the hands and face, can be made without even having to first make soap. Here's how:

Grate two bars of glycerine or unscented soap. Melt gently over hot water in a double boiler. When liquid, add one oz. finely chopped herbs, (sweet marjoram for example) a few drops essential oil, and one tablespoon fine oatmeal or bran, and pour into moulds.

HERB PILLOWS

Herbs that keep their scent when dried have been used for centuries in bedding. Favorites include woodruff, lavender, rose petals, camomile and hops. Hops have long been used in pillows as a remedy for insomnia. The herbs can be put in their own pillow case or a sachet can be added to a regular pillow case.

SOOTHING HERBAL PILLOW

    2 parts dried mint
    2 parts dried camomile
    1 part dried hops

PET PILLOW TO PREVENT FLEAS

Cats . . . equal parts dried camomile and pennyroyal.
Dogs . . . equal parts dried rue and costmary.

HERBAL COSMETICS

For centuries, cosmetics were simple herbal recipes made in the home from garden herbs. Just in the last few decades has the mass-production of cosmetics become an industry.

The two major advantages you will have in selling herbal cosmetics are that you can offer pure and natural ingredients and that you can adjust the blending of those ingredients to suit the exact requirements of each customer. Just as in cooking, the herbs you use can be main ingredients or flavorings. Here are just a few of the many common herbs used in cosmetics together with their general characteristics.

Borage . . . leaves are softening and cleansing.
Camomile . . . flowers are soothing, cleansing and gently astringent.
Chervil . . . leaves are gently astringent.
Comfrey . . . leaves are softening and healing.
Fennel . . . leaves are cleansing and gently astringent.
Lavender . . . antiseptic.
Lemon Balm . . . leaves are soothing and astringent.
Lovage . . . leaves and roots are cleansing and deodorant.
Mint . . . leaves are stimulating, antiseptic and healing.
Rosemary . . . leaves are invigorating and astringent.
Thyme . . . leaves are deodorant and anticeptic.

HERBAL BATH BAGS

The simplest way to use herbs to soften or tone the skin is to use them in the bath. Like a sachet, the herbs are sewn up in small muslin bags before use. Just add the bag to the bath before running the hot water, squeezing occassionally to fully extract the juices. The bags can be re-used several times.

RELAXING BATH BAG
Equal parts lavender and bay leaves.

BATH BAG FOR YOUTHFUL SKIN

Equal parts lavender and rose petals.

HERBAL SKIN TONIC

Equal parts comfrey and mint.

STIMULATING BATH BAG

Equal parts marigold and mint.

HERBAL BATH FOR MEN

Equal parts mint and thyme.

HERBAL FOOT BATH

Equal parts sage, thyme, sweet marjoram and bay leaves. Put the bath bag and one tablespoon of salt in a large bowl of boiling water. Steep for at least five minutes before soaking feet. Just the solution for tired feet.

HERBAL SHAMPOO

A simple herbal shampoo can be made using a favorite herb. Mix equal parts pure baby shampoo and herb infusion. A herb infusion is simply a strong tea made by pouring boiling water over herbs, which is left to infuse in a covered pot for at least one hour before straining. For the infusion, use two oz. dried herb or four oz. fresh herb to one pint boiling water. Do NOT use an aluminum pan, as it may react with the herbs and taint the infusion.

HERBAL HAIR RINSE

Vinegar is one of the best hair rinses for all types of hair. It helps restore the natural acid balance of the scalp, stops itching and helps control dandruff. It cleanses the hair and scalp. Mix up a herb vinegar as described in the section on culinary herbs. Herbs that are especially good for vinegar hair rinses include:
Camomile . . . the flowers soften and lighten hair.
Rosemary . . . perhaps the best herbal tonic and conditioner; giving luster and body to the hair.
Sage . . . leaves are also tonic and conditioning; and darken hair slightly.
A blend of rosemary & sage are helpful in controlling dandruff.

RECOMMENDED READING
RODALE HERB BOOK

Mentioned in the last section, this book has much practical information and formulas on the fragrant herbs and products.

POTPOURRI, INCENSE AND OTHER FRAGRANT CONCOCTIONS

Ann Fettner 1977. Here is the how-to of drying flowers and herbs, mixing them and designing the containers for sachets, incense, candles and perfumes.

THE FRAGRANT GARDEN

Louise B. Wilder. This Dover reprint of the 1932 classic contains much hard-to-find information about hundreds of fragrant plants.

GROWING HERBS AS AROMATICS

Roy Genders. The art of growing and using aromatic herbs for pomanders, potpourris, scented waters, rose perfumes and other uses.

All three books are available from: Richters, Goodwood, Ontario, Canada L0C 1A0

FRAGRANT HERB SOURCES

The Fragrant Path, P.O. Box 328, Fort Calhoun, NE 68023
This delightful catalog is devoted to fragrant plants of all kinds, including many herbs. Send $1 for latest issue.

Richters, *specializes in herbs, and their eighty page catalog is an education on the subject. Plant descriptions include uses for the plants, and the herb book selection is the most comprehensive of any source. Send $3 for current issue. Richters, Goodwood, Ontario, Canada L0C 1A0*

Sandy Mush Herb Nursery, *Rt. 2, Surrett Cove Rd., Leicester, NC 28748, carries over 700 herb varieties, including rare and hard-to-find plants and many varieties of scented geraniums. Catalog is $4.*

MEDICINAL HERBS

Since the beginning of recorded time, medical treatment has been based on herbal remedies. As modern medicine has developed, thousands of plant species have been tested and utilized in today's vast array of drugs. For example, foxglove is a natural source of digitalis, used in treating heart disease. In most instances, laboratory testing confirms the traditional healing claims for herbs. This has led to a revival in the use of herbal remedies, and to a growing demand for dependable sources of supply for these herbs.

Drug manufacturers have also found that the use of natural ingredients is generally more cost-effective than the use of synthetic ingredients, and preferred by their customers.

Because of space limitations, only those medicinal herbs with proven demand and an established wholesale distribution network are listed below.

Remember that if herbs are grown with intent to use for medicinal purposes, they fall under the drug guidelines of the Food and Drug Administration. So be sure to seek professional advice before selling medicinal herbs.

BORAGE

Borage grows well in a wide range of soils. Space plants two feet apart in full sun. Borage should be harvested when it begins to flower. Dry the flowering tops. Borage will yield an average of 1000 pounds to the acre. Average wholesale price in 100 to 500 pound quantity is $2 per pound, with the average retail price at $8 per pound.

CALAMUS

Calamus, sometimes called Sweet Flag, grows wild in marshy areas, but adapts well to rich moist loam. During the Depression, calamus was chewed as a tobacco substitute! The roots are dug at harvest time and dried, yielding an average 2,000 pounds per acre of dried root. Average wholesale price for 100 pound quantities — $2.50 per pound, while it retails for $9-$10 per pound.

CATNIP

This hardy perennial is easy to grow, and in great demand, not just for pet toys, but for it's medicinal properties. Many growers sell the flowering tops to the botanical wholesalers, and the stems and leaves to pet supply wholesalers. Yield averages 1500-2000 pounds per acre. Wholesale prices average $2 per pound in 500 pound lots, with retail at $9 per pound.

ECHINACEA

Echinacea, also known as purple cone-flower, was used by the Plains Indians to treat a wide variety of ailments. Much of the current supply comes from foragers, who dig the mature roots in the wild. There is a real need for commercial growers of echinacea, because the increased demand is threatening to eliminate the wild population. Keep in mind through that plants started from seed will need from three to four years before commercial sized roots can be harvested. Current prices make the wait worthwhile, with 100 pound quantities wholesaling at over $8 per pound, and retailing for over $20 per pound.

GINSENG

American ginseng is an ordinary-looking plant which grows on the shaded forest floor. It's value lies buried in the slow-growing tuberous rootstock.

The Chinese have valued the root for thousands of years as the most potent of medicinal herbs and a regenerative tonic. Since 1717, when it was discovered here in the American colonies, ginseng has been exported to the Orient. Today the Orientals living in the Pacific basin – Hong Kong, Japan, Taiwan, Malasia, Singpore and the Philippines consume 85% of the American crop. The price per pound has gone from 7 cents in 1717 to $155 in 1987.

There are about a dozen major ginseng exporters in the U.S., with a network of agent-buyers who will quote the ginseng farmer a price. Buy bids can be obtained by mail on the basis of a sample but once a grower establishes himself as a producer of quality roots, buyers will often come to the farm to grade and offer a bid on the roots as they are drying.

According to Dr. Tom Konsler, professor at North Carolina's Horticultural Crop Research Station and an authority on ginseng, "American ginseng has great potential as a small-scale cash crop." But he cautions, "Ginseng production is not a quick or easy way to get rich. By it's nature, it requires great patience."

Growing ginseng means duplicating it's native forest environment, and there are three basic ways of doing this. Most ginseng today is grown under artificial shade. It is becoming a less profitable method because the root price is lower than other methods (currently $30 per pound) and the investment in the shade structure and other production costs are quite high. Another approach, the wild-simulated production method, is the easiest, the least expensive, and the slowest. Prices are much higher, currently $150 a pound, but your first harvest is eight years away. One person can plant an acre per year, even on steep hillsides and ravines. Because the seeds are scattered and left to grow naturally, the roots look like wild roots and bring the highest prices.

The third method is called woods-cultivated and involves preparing growing beds in your woodland. As with the wild-simulated approach, the main expense is labor. Yields are higher, but prices average just $60 per pound. With just a tiller and hand tools, one person can plant a small area of an eighth acre per year for a sustained income starting in six years.

As wild ginseng becomes scarcer, demand for woods-cultivated and wild-simulated ginseng should increase to help fill that high-priced end of the market. So the long-term outlook appears excellent.

Until recently, it has been extremely difficult for anyone interested in getting started in ginseng to find reliable information on the practical growing details. But now a new book is available that is practical, thorough and readable. It even contains a list of established buyers and suppliers. The book is available directly from the author, W. Scott Persons, who is also a grower & supplier of seed and seedling planting stock. For current price, write him at: Tuckasegee Valley Ginseng, Box 236-G, Tuckasegee, NC 28787.

GOLDENSEAL

Goldenseal is a slow-growing perennial which requires, like ginseng, four to five years between planting and harvest. American Indians used goldenseal for a wide variety of ailments, and it has become quite popular in the past few years. Shading is required to duplicate natural forest growing conditions. You can expect 1500 to 2500 pounds to an acre, with current wholesale prices averaging $18 per pound, and retail prices over $60 per pound.

HOREHOUND

This hardy perennial native of Europe is tolerant of a wide range of growing conditions. Start from seeds or root division. The leaves and flowering tops are harvested in full bloom, and should yield about 1500 pounds to the acre. Wholesale prices run from $1 to $1.50 per pound, retail about $6 per pound.

LOBELIA

Lobelia is an annual native to North America. It contains an alkaloid, lobeline, which acts as a depressant to the central nervous system. This is used in many over-the-counter "stop smoking" products. An acre can produce 1500 pounds, and the current price is $2 to $3 per pound in 100 pound quantities and up to $12 per pound retail.

MUGWORT

Mugwort was named for it's traditional use as a hop substitute in beer making. It's a hardy perennial that likes a sunny spot and rich soil. The roots are harvested in the fall, then washed and dried. Between 800 and 2000 pounds of root can be harvested from an acre, at current prices of $1 to $2 per pound wholesale and $12 retail.

MULLEIN

This biennial herb is a common fencerow weed in most parts of North America. It prefers terrible soil and dry conditions! As a botanical, or medicinal herb, it is used in ear drops, eye washes and as a tea. In 100 pound to 500 pound quantities it wholesales at $1 to $2 a pound, retailing at $7 a pound. Per-acre production figures are not now available for mullein, as most of the current supply is foraged.

PENNYROYAL

European pennyroyal is a perennial member of the mint family. It prefers a sunny spot with rich sandy loam. Propagate by dividing root runners in spring. Pennyroyal is used in teas and in insect repellants. An average acre will yield about 1200 pounds, at a wholesale price of $2 per pound and $8 per pound retail.

ROSEHIPS

The rugosa rose is a hardy shrub which does well in most growing situations. At harvest time the buds resemble cherry tomatoes and are a rich source of vitamin C, (as much as 60 times richer than oranges) and the petals also can be harvested for their oil. Yields are variable, and current wholesale prices are $1 per pound and $5 per pound retail.

VALERIAN

Valerian is known as the "tranquil herb", because the root acts as a central nervous system depressant. Thus, like nicotine, it will stimulate you when you're tired and calm you when you're excited. The plants are best started by division in the fall, as the seeds do not germinate well. The roots are harvested in the fall, washed and dried. An acre of plants will produce almost a ton of roots, with current prices ranging from $1 to $2 per pound wholesale and $10 per pound retail.

SELLING MEDICINAL HERBS

Since most medicinal herbs are processed into a manufactured product such as salves or pills, a small grower should plan to market their crop through wholesale buyers. These include regional wholesalers, cooperatives, manufacturers, exporters and brokers. Selling to wholesale buyers allows you to spend more time on growing and less on marketing.

To begin selling your crop to the wholesale buyers, write to them, asking about their current needs, prices and grading requirements. Then send out a simple one page flyer or letter listing what you have available, prices, and terms of sale. Most growers, after getting established, will be able to sign contracts for all they can grow. Consider storing part of your crop to sell during the off-season, when prices are usually higher.

The markets for medicinal herbs are changing constantly, and a buyer that is paying premium prices one year may be out of business next year. The best way to find a market for your crop (ideally before you even plant the seed) is to stay in touch through herb trade groups and newsletters. They can help you find out what is in demand, how much it's selling for and who's buying it. Here's a list of resources to help you.

American Herb Association
P.O. Box 353, Rescue, CA 95672

American Herbal Products Association
215 Classic Court, Rohnert Park, CA 94928

The Business of Herbs
Northwind Farm, Rt. 2, Box 246, Shevlin, MN 56676

Chemical Marketing Reporter
Schnell Publishing, 100 Church St., New York, NY 10007
Publishes the latest information on processed herb prices and dealers. Their annual lists all current dealers.

International Herb Growers and Marketers Association
P.O. Box 281, Silver Springs, PA 17575
This trade group publishes a newsletter, publishes a directory of growers and buyers, organizes special workshops and an annual conference, covering topics like Marketing Botanicals, The State of Herbal Medicine, and Hot New Botanicals. If you miss a conference, cassettes are available on every topic.

RECOMMENDED READING

HERBAL BOUNTY, by Steven Foster, covers all aspects of eighty common herbs. His extensive background as a commercial-scale grower of herbs adds a great deal to this thorough book.

Must reading for any serious grower of herbs is THE POTEN-TIAL OF HERBS AS A CASH CROP, by Richard Alan

Miller. The book covers plant requirements, propagation methods, drying and storage methods, processing and marketing. Available through Acres USA, P.O. Box 9547, Kansas City, MO 64133.

Paul Hsu, a Wisconsin exporter of ginseng, has a variety of books and leaflets on ginseng. He also sells seed and rootlets, buys mature roots for export and is most helpful with beginners. Write to him at: Hsu's Ginseng Enterprises, P.O. Box 509, Wausau, WI 54402.

There are two newsletters that are a must-read for medicinal herb growers. The first, **Herbalgram**, (P.O. Box 12006, Austin, TX 78711) includes a market report in every issue, in-depth articles on individual herbs, and news of the industry, all packed into 28 informative pages.

The second, **The Herb, Spice, and Medicinal Plant Digest,** (Dept. of Plant & Soil Sciences, Stockbridge Hall, University of Mass., Amherst, MA 01003) is published under the auspices of the U.S.D.A.'s Cooperative Extension Service. It covers production techniques, chemistry of plants, new developments such as natural herbicides from plants and reviews new literature that would be helpful to growers.

SEED AND PLANT SOURCES

In addition to the sources listed under culinary herbs, the following suppliers should be helpful.

Casa Yerba
3459 Day's Creek Rd., Days Creek, OR 97429
($1 for catalog)

Gilberties Greenhouse
Sylvan Avenue, Westport, CT 06880
($1 for catalog)

Meadowbrook Herb Gardens
Route 128, Wyoming, RI 02898
($1 for catalog)

Otto Richter & Sons, Ltd.
Goodwood, Ontario, CANADA L0C 1A0
($3 for catalog)

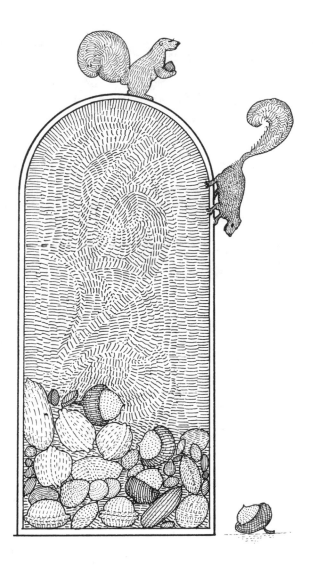

NUT CROPS

A few years ago, Bill Schildgen bought six acres of sagebrush and rocks in eastern Washington state with the idea of creating a natural grove of nut trees. "They told me I was crazy to try to grow nut trees here," he said. But now his neighbors are buying nut trees from him! His 3 1/2 acre grove resembles a natural forest. Walnuts and Chinese Chestnuts, planted on sixty foot centers, form the upper story of the grove. Filberts, which grow well in partial shade, fill in the twenty foot high second story.

In addition, peach, plum and pear trees are planted in the sunny corners and edges of the grove. Table and wine grapes are trellised around the edge of the grove. Bees from his own hives do the pollinating, and peafowl patrol for insects.

The grove is just reaching it's prime, producing eight thousand pounds of nuts, worth $12,000 each year. But the nut and fruit tree grove is only part of Bill's cash crop income. On the remaining 2 1/2 acres, he grows seedlings and grafted trees. His trees are of such high quality they're sold out for the next two years, and several nursery/seed companies have standing orders for his nuts. His income from seedlings and grafted trees alone is $20,000 per year. Bill says this grove could produce for another fifty years, with very little further work.

Bill Schildgen is a fine example of what is possible with a well-managed small nut orchard. In this section on nuts, we've listed more information on Almonds, Chestnuts, Filberts, Pecans and Walnuts. These nut varieties are the most promising for small growers because they are already widely grown in commercial quatities and have established markets. There are many other nut varieties with commercial potential, such as macadamia nuts, jojoba, pistachios and cashews, all of which have quite exacting growing requirements. If one of these varieties is being grown in your area, you may want to consider planting it.

ALMONDS

The United States is the leading almond producing country in the world, with about half of the total crop. Much of the U.S.

crop is grown in the Sacramento and San Juaquin valleys of California, Texas, and other areas of the Southwest. This is due to the almond's requirement for mild winters and a long hot growing season. Almonds are difficult to grow in the eastern U.S., due to higher than ideal humidity.

The almond produces a vigorous, medium sized tree, usually 20 to 30 feet tall. The tree may begin to bear as soon as the third year, and should be producing a good crop by the sixth or seventh year. The yield of a mature orchard typically ranges from 1500 to 3000 shelled pounds per acre per year.

CHESTNUTS

The American Chestnut was once the dominant tree in hardwood forests until the chestnut blight fungus destroyed it early in this century. In recent years, most commercial orchards have used the blight-resistant Chinese Chestnut. Even with many new orchards being established each year, the U.S. still imports millions of dollars worth of European chestnuts each year to satisfy the demand.

Chinese chestnut bears at a younger age than most nut trees, usually three to five years after planting. They are regular bearers compared to other nut trees, which tend to have light and heavy bearing years.

Researchers have developed a hybrid chestnut that combines the best qualities of American and Chinese chestnuts, and can be grown in most areas of the U.S. This hybrid chestnut, called the "Dunstan", was first developed in the 1950's and has since been extensively planted with good results. For more information, write to: Chestnut Hill Nursery, Route 1, Box 341, Alachua, FL 32615.

Two big pluses for chestnuts are that they can be grown on land too hilly or poor for other crops, and the timber value at maturity is similar to black walnut. Average orchard yields are 2000 to 3000 pounds of nuts per acre.

FILBERTS

The filbert, sometimes called the hazelnut, was one of the first shrubs to appear following the receding glaciers of the last ice age. From 8000 B.C. to 5500 B.C., the filbert was the dominant vegetation of northern Europe.

Today, filberts are commercially cultivated in areas of the world that have a climate influenced by a large body of water, resulting in cool summers and mild winters. For example, about two-thirds of the total world filbert production comes from small groves in Turkey, located along the southern coast of the Black Sea.

Most of the U.S. crop is grown in western Oregon and Washington. Current production is not enough to meet the demand, with half the demand imported each year. Filberts have two markets, in-shell and kernel. The in-shell market is seasonal, primarily the holiday season from Halloween through Christmas. Kernels are sold to bakers, candy-makers and salters (for mixed-nut packs).

The major cultivar in the U.S., the Barcelona, is larger than European varieties, and commands a premium price in the export market to Europe where they are used in chocolate candy.

In the U.S., filberts are usually grown as a single-trunk tree, reaching twenty to thirty feet in height. They begin to bear early, but commercial production does not begin until the sixth or seventh year. A good orchard will produce an average of 1500 pounds of dried nuts per acre.

Filberts do tend to bear a heavy crop one year and a light one the next. A well-managed orchard will remain productive for forty years. One tree in Scottsburg, Oregon was planted by an English sailor in 1858!

"In a world where some consider that faster is better and all change is good, there is a certain satisfaction in planting, grafting and nurturing trees that will produce food for years to come."

PECANS

The first French and Spanish settlers in America found native pecan trees growing in the valleys of the Mississippi River. Today the pecan is commercially grown in much of the South, and in Texas, New Mexico, Arizona and California. Climate controls where pecans can be successfully grown, and includes a long frost-free growing season, warm summers and adequate moisture.

Most pecan cultivars will begin to bear lightly at four years, with commercial production at eight years. As the tree continues to mature, nut production will continue to increase up to twenty years of age. A mature orchard will typically yield 1200 to 2000 pounds of nuts per acre. Only about ten to fifteen percent of pecan production is sold in-shell. Most is sold shelled to bakeries, candy makers and dairies (for ice-cream).

WALNUTS

The two areas of interest to commercial walnut growers are nuts and timber. Commercial nut production is limited to the "English" walnut, which gained the name only because it was first brought to this country in English ships. The alternate name of "Persian" walnut is more accurate as an indication of the source of the trees.

For nut production, California accounts for over half of the world production. But as subdivisions continue to displace walnut orchards in California, other parts of the country are increasing commercial plantings, as the Persian walnut can do well in a wide variety of climates. A mature orchard takes eight to ten years to come into production, but can produce up to six thousand pounds per acre.

If you live in an area not suited for growing nuts, consider black walnut for timber. During the seventeenth century, walnut became the preferred wood for furniture. The reign of Queen Anne (1702-1714) is known as the age of walnut because of it's popularity. Walnut has remained king of the woods ever since.

Shortly after the Civil War, the Singer Sewing Machine Company sent "timber lookers" into the South to scout for walnut. They bought big standing walnut trees, paid in advance, branded the trees and hired guards! But by 1880 walnut was getting scarce, and the Singer folks began to use other woods, making bird's eye maple the new wood fashion.

Today, black walnut is even scarcer. Single trees have been sold for $20,000, and the timber rustlers found that with only a chain saw and a truck they could steal from suburban front yards. In the suburbs of Chicago, the walnut bandits have a reputation worse than dutch elm disease!

One good example of the profit potential of walnut is gunstock blanks. Cull or reject logs are cut into three foot lengths and dried. To prevent checking and splitting while drying out, these short logs are soaked in PEG (polyethelene glycol) and then dried. Because of the demand for fancy walnut gunstocks, the market is brisk, with demand usually greater than supply and prices of up to $300 per blank.

To utilize even smaller sections of cull logs, such as limbs and crotches, manufacturers are buying reject pieces and using them for novelties such as pen holders, paperweights and bookends. For these uses, rough knotty wood is just as good, and in some cases better, than the high-grade logs.

Gene Garrett, a University of Missouri forester, has developed a program of double-cropping black walnut trees with pasture crops for harvesting or livestock grazing. He plants the trees in 40 x 10 foot rows, or 108 trees per acre, and plants the other crop in between the rows. Up to seven feet is left between the trees and the crop to avoid damage to the tree's root system. Intercrops of high-value crops such as raspberries and blueberries work too.

Garrett says this plan provides income four different ways. For the first few years the only income is from the crop planted between the trees. As the trees become larger, they are thinned from 108 to 27 trees per acre. Wood from these early thinnings can be sold to sawmills. Woodworking hobbyists pay premium prices for wood too small for the sawmill.

After a few years, the walnut trees begin to produce nuts for harvesting. When the thinned walnut trees get to be 40-60 years old, they are harvested for veneer logs. High quality veneer logs now sell for up to $3,000 each.

According to Bruce Thompson, author of "Black Walnut for Profit", a planting established today on a suitable site could bring $100,000 per acre in timber value alone in just 35 years! In addition, Mr. Thompson says "If the planting is done where a substantial population increase occurs during the 35 intervening years, the projected values might easily double due to skyrocketing land values."

MARKETING NUT CROPS

To determine current wholesale (bulk) prices and buyers for your crop, contact the grower's associations listed below. If none are listed for your area, your county extension agent can probably refer you to local sources.

Alabama Pecan Growers Association
Route 2, Box 227, Daphne, AL 36526

California Almond Growers Exchange
P.O. Box 1768, Sacramento, CA 95808

Federated Pecan Growers of U.S.
Warren Meadows, L.S.U. Campus
P.O. Drawer AX, Baton Rouge, LA 70803

Florida Pecan Growers Association
Tim Crocker, 199 McCarty Hall
University of Florida, Gainesville, FL 32611

Georgia Pecan Growers Association
P.O. Box 1209, Tifton, GA 31794

Mississippi Pecan Growers Association
Box 5426, M.S.U., MS 39762

Northern Nut Growers Association
9870 S. Palmer Road, New Carlisle, OH 45344

Nut Growers of Oregon & Washington
P.O. Box 23126, Tigard, OR 97223

Other wholesale outlets for your nut crop include your local food stores, and "gift packs" wholesaled to your local shops.

RETAIL NUT MARKETING

Your nut crops can be retailed at local or regional farmer's markets, your own roadside stand, direct from the tree (U-pick), or in bulk to local food co-ops (semi-retail).

Many growers advertise their regional specialties in national publications in-season and are pleasantly surprised at the number of folks willing to go out of their way to get specialty crops not readily available at local markets. An additional source of income is "value-added" nut products such as nut butters, candies and cookies which can be produced during off-times.

Growers who enjoy nursery work may wish to consider selling nut tree seedlings and grafted varieties from your own orchard. Many growers find this to be one of the most profitable areas to develop.

NUT CROPS
RECOMMENDED READING

The "Bible" for nut tree growers is titled "Nut Tree Culture in North America". This 466 page encyclopedia, with contributions by 25 specialists, is probably the most complete book on North American nut trees ever written. 29 chapters cover all common nut varieties, breeding, propagation, pruning, disease and insect control, and nut production. For a current price, write to the Northern Nut Growers Association, Broken Arrow Road, Hamden, CT 06518.

The Northern Nut Growers Association, mentioned above, welcomes new members. They publish an annual (the last one was 26 articles and 172 pages), and issues a quarterly newsletter titled "The Nutshell". Write to Ken Bauman, Treasurer, for current dues, at : 9870 S. Palmer Road, New Carlisle, OH 45344.

The U.S. Department of Agriculture, Publications Division, publishes a wide variety of titles and can also refer you to other departments for even more specialized information. Write: Office of Governmental & Public Affairs, Publications Office, USDA, Room 114A, Washington, DC 20250.

For those interested in growing the Persian walnuts, the booklet "Establishing a New Walnut Orchard" (No. 21157) is available through the Cooperative Extension Service, USDA, University of California, Berkeley, CA 94720.

Serious growers/planters of timber walnut should consider joining the Walnut Council, at P.O. Box 41121, Indianapolis, IN 46241. They provide a quarterly magazine, an annual summer get-together which includes field tours of established plantings, gunstock mills, veneer mills, and nut processors.

NUT CROPS
SEED & PLANT SOURCES

There are thousands of nurseries and seed companies in North America. Trees can be "site-sensitive", so it is important to purchase your planting stock from a local source whenever possible. This also gives you the opportunity to chat with the nursery staff personally, and get specific recommendations about the varieties you choose. So before ordering by mail, check with your county extension agent for local sources.

Forest Tree Seed Orchards
U.S. Forest Service, USDA, Washington, D.C. 20250

A directory of industry, state and federal forest tree orchards in U.S.

Nursery Source Guide
Brooklyn Botanic Gardens, 100 Washington Avenue, Brooklyn, NY 11225

This $2 guide lists major retail and wholesale nurseries in the U.S. and includes descriptions of over 1200 trees and shrubs referenced to suppliers.

Bear Creek Farms
P.O. Box 248, Northport, WA 99157

Cold-hardy nuts and grafted trees for nothern growers. Send $1 for catalog.

Burnt Ridge Nursery
432 Burnt Ridge Road, Onalaska, WA 98570

A wide variety of grafted filberts, persian and black walnuts, chinese and hybrid chestnuts.

Chestnut Hill Nursery
Route 1, Box 341, Alachua, FL 32615

Only source for "Dunstan" blight-resistant hybrid chestnut.

Earl Douglas Chestnut Seeds
R.D. 1, Box 38, Red Creek, NY 13143

A source for blight-resistant American chestnut seed.

Grimo Nut Nurseries
R.R. #3, Lakeshore Rd, Niagara-on-the-Lake, Ontario L0S 1J0

Specialist in northern nut varieties.

Nolin River Nut Tree Nursery
797 Port Wooden Rd., Upton, KY 42784

Specialists in grafted nut trees — pecans, hicans, hickories, butternuts, walnuts, chestnuts — over 100 varieties.

Raintree Nursery
265 Butts Road, Morton, WA 98356

Specializing in varieties proven in the maritime northwest.

Saint Lawrence Nurseries
RD No. 2, Star Route 56-A, Potsdam, NY 13676

Cold hardy varieties of many nut trees.

LANDSCAPING PLANTS

Deep in the mountains of North Carolina, a semi-retired tabacco farmer, Spencer Dellinger, heard about a new program at the local Ag university to help farmers shift into "high-value" crops. The local extension agent said "The market's there . . . all you need to do is grow the plants."

After checking several possibilities, he decided to grow container ornamentals for retail sales. Now, after five years, his six acres contain thousands of seedlings, container and fieldgrown plants and landscaping trees. Spencer has focused on azaleas, rhododendrons, tiger lilies, Japanese and Crimson Glory maples, Frazer firs, Norway and Colorado blue spruce and Andorra juniper.

Each year his stock is completely sold out, (without any paid advertising!) 80 percent of his plants are sold to local residents who appreciate quality plants at a reasonable price. The rest goes to landscapers and two nearby retail nurseries.

He charges one price to everyone, regardless of quantity. Prices range from $8 to $10 a plant, with production costs at about $2 per plant. Since getting established, he can simply take cuttings or seed from the "mother" plants, rather than buying stock.

As for profits, he will only admit he's finally making more money than ever before in his life. If he wanted to, he could just grow out his existing stock of 80,000 plants and retire, set for the rest of his life!

Like Spencer Dellinger, you will have to make important decisions in the beginning about what to grow, how best to grow it, and how to sell what you grow.

There are three basic commercial growing methods used: most plant nurseries tend to specialize in one of them. Each method (field-grown, container-grown, and bedding plants) is covered here, with basic information to help you choose based on your own local markets and interests.

FIELD-GROWN PLANTS

Perhaps you've decided to plant a field of scarlet maples or andorra junipers for future sales. First, take time to check out local markets. Will you grow nursery stock for retail garden and nursery centers? They have a fairly short season and prefer to sell small plants that will fit in the trunk of a car. The most popular plants for retail nursery centers are: pines, spruces, yews, junipers, arborvitae and flowering shrubs.

If you prefer to sell to landscapers, you will find they need larger plants over a longer season, but you will have to do more work digging larger and heavier stock. For example, a landscaper sized shade tree could require a large root ball weighing over 100 pounds.

You may wish to stay small, specializing in plants that are rare or difficult to grow. With thousands of plants to choose from, finding a "niche" in the marketplace should not be a problem. Many small specialty nurseries are able to serve a broader regional market this way, focusing on exotica such as Japanese maples or bamboo or cactus.

If you have an acre or less, you should probably focus on the retail market, and emphasize a diversity of plants. Remember that you will then be competing with other retail outlets, and offer your customers something extra to keep them coming back year after year.

Check with your county extension agent for the name of the nearest nursery specialist on the extension service payroll. They are usually quite expert on a wide variety of topics unique to your new business, from fertilizing to pest control.

Before you buy a single plant, talk to at least a half-dozen retail nurseries in your area. Tell them you're planning a wholesale nursery and ask what plants they are selling and what they would prefer to buy from local sources.

It is very important to select plants that are suited to your location. For example, if the demand is sufficient and your soil is light and sandy, you may want to specialize in rhododendrons and azaleas. If you have heavy clay soil, you won't be able to grow them easily. Similarly, if you live in the North, buy starts from northern growers. Otherwise, you may get winterkills and branches may die back in winter.

Remember that you'll have to do most of your planting and harvesting in Spring. Weather that is perfect for planting also excites customers who can't wait to landscape their own yards!

With good advertising and customer education, the fall season can be an excellent time for selling landscaping plants. However, since many homeowners are burned out after six months of mowing and trimming, you'll have to entice them with tempting sale prices.

Here are the **"TEN COMMANDMENTS"** for a successful plant nursery.

1. Pick your suppliers carefully to insure disease and insect-free stock with a high survival rate.

2. Check to insure that the stock that you order will be hardy in your local area. Determine the "hardiness zone" before you order.

3. Match your planting stock to your planting site. For example, junipers need full sun, yews don't like wet feet, willows do.

4. Order transplants over seedlings when possible. Each time a plant is transplanted, the root system becomes denser and hardier. The extra expense will be repaid in superior growth and plant survival rate.

5. Plant early. By starting as soon as your land is thawed, the root system can "settle-in" before the new upper growth begins. Never allow roots to dry out when planting and water regularly.

6. Allow room to grow. With adequate spacing, your stock will fill out evenly. No one wants a one-sided plant!

7. Cultivate and weed. This prevents weeds from choking out the lower growth on your plants and competing for available water and nutrients. Cultivating also reduces surface roots, making transplanting easier.

8. Use organic or slow-release fertilizers. The small extra expense will be amply repaid in higher profits when you sell your larger and healthier plants.

9. Prune often to remove dead growth and encourage new growth.

10. Check your plants often to catch insects or diseases before they get out of control.

CONTAINER GROWING

In recent years there has been an increase in the number of plant nurseries using container growing. Container growing offers several advantages over field growing:

1. The number of plants per acre can be much greater.

2. Container-grown plants can be sold year-round, regardless of weather.

3. Container-grown plants can be grown faster because

the soil can be blended to suit a particular plant.

4. Container growing is ideal for varieties that are difficult to transplant, such as pyracantha.

5. You will have saleable plants sooner.

WHICH PLANTS TO GROW

Just like field-grown plants, your choice of varieties will be controlled by the lowest cold temperatures in your area. Order a copy of the U.S. Department of Agriculture plant hardiness zone map (misc. pub. No. 814) The hardiness of plants designated in most nursery catalogs uses the zone numbers given on this map.

To learn which plants are in demand visit wholesale and retail nurseries locally to learn what is selling well. Also, try to include new varieties that are not yet widely available in your area. Commercial rose growers, the largest segment of the nursery industry in dollar volume, never just sell roses; they are always selling new varieties.

Start out with a limited number of varieties; specializing in certain plants. For example, you could specialize in container-grown perennial herb plants, offering half a dozen varieties of such culinary herbs as mint, rosemary, sage, and thyme for resale to nursery centers and your own customers.

WHAT SIZES TO GROW

The smallest container usually used for container growing is the one gallon size. For many plants such as ground covers, this will be the only size can necessary, regardless of the plant size. When you grow the larger plants and shrubs, which require larger containers, you will usually have much less competition, and the higher prices will make up for the extra time required to grow larger stock. Also, the larger sizes produce a more beautiful plant.

You will need to make a choice: smaller plants for the volume lower priced market or larger plants for the higher priced markets. Most purchasers associate quality with size, so well-shaped plants only eight inches high are not nearly as attractive as the same plant twelve inches high. Each may be the best quality, yet size will make the difference!

Also keep in mind that to do the same dollar volume, you will have to grow many, many more one gallon plants for sale than the larger sizes, such as three and five gallons. The small nursery, particularly retail, should focus on quality larger plants which are hard to find at most outlets that depend on volume. This reduces the labor involved in selling considerably.

SELLING YOUR PLANTS

You should begin selling long before your plants will be ready if you are wholesaling your plants. Visit retail nurseries, landscapers and stores and tell them about your plans, the plants you are planning to grow, and the quantities and sizes you'll be offering. Ask them for suggestions and advice! You must remember that in addition to growing what the customer wants, you have to let them know you have it, and persuade them to buy it from you. Invite your prospective customers to your nursery; make sure it is neat and well-maintained. Once they visit, the ice is broken and they will be ready to buy from you.

BEDDING PLANTS

Some nurseries specialize in starting plants from seed or cuttings for resale to other nurseries or to the public. These are called liners or bedding plants. Most bedding plants are produced in a greenhouse or coldframe. Although there are growers who specialize in producing liners such as Christmas tree seedlings, a fairly large equipment investment and specialized knowledge is required. For the beginner, it's best to start small; growing flowers, vegetables and herbs. Quality and reliability are the key to success.

You must produce quality plants to sell them, and your plants must be true to label and perform well when planted. In addition, your plants must be ready to deliver or sell at the right time.

In flowers, there are annuals that are popular every year: marigolds, petunias, begonias, geraniums and salvias. Most are sold in multiple containers at a lower price than individual plants. But some customers are willing to pay for larger plants in individual containers. They are also willing to pay extra for unusual bedding plants that the marigold-petunia customer might not be able to afford.

The outdoor planting season for most annuals is four to six weeks in most areas. To have plants in prime condition through the entire planting season it is necessary to make successive plantings. Plants that require from four to six weeks to be ready to sell include calendulas, cosmos, marigolds and zinnias. From six to eight weeks is necessary for celosias, phlox, dahlias, verbenas, eggplant, peppers and tomatoes. From seven to nine weeks is needed for snapdragons, petunias, sweet alyssums and vinca rosea.

Petunias are at the top of the list of the most important bedding plants, with marigolds a close second. Mid-January to mid-February is the best time to start petunia plants. Lots of petunias always mean good profits for the bedding plant grower.

Many homeowners like to get a headstart with larger and more colorful bedding plants. So they are willing to pay extra for geraniums, fuschias, begonias, petunias and impatiens in three to six inch pots that are blooming when purchased. Geraniums remain the most popular plant sold in larger pots. Fuschias are next in large pot sales, both the upright and trailing variety.

Late in the season, there will be demand for zinnias, marigolds and salvias, which will do well in summer heat. They can be used to fill in bare spots and provide instant color. Folks who like to keep fresh-cut flowers in their home will buy the larger plants late in the season.

Many bedding plant growers start the season with hardy vegetable plants such as cabbage, lettuce, broccoli, cauliflower and brussel sprouts. Another interesting sideline is smaller perennials such as chrysanthemums in peat pots or small containers. They will sell much better with colorful labels giving growing information. This is especially important for plants not in flower.

ORNAMENTAL GRASSES

The ornamental grasses are enjoying a well-deserved revival in popularity today. Landscape designers are re-discovering them and nurseries are stocking a great diversity of plants and seeds. These undemanding perennials add their striking form and color to the landscape for much longer than flowers, some even lasting through the winter dormant season.

Pampas grass, the "queen of ornamental grasses" is what most people immediately think of when ornamental grasses are mentioned. Yet there are dozens of other ornamental grasses, most of them hardy in the North. In addition to the perennial grasses, there are about 25 annuals which are grown from seed and have decorative flowers than can be dried for boquets. For more information on these, see chapter 10.

The perennial grasses are easy to grow and maintain, and are bothered by few insects or diseases. You can propagate most of them by dividing the root clump as it matures and enlarges. Most ornamental grasses fall into one of these groups, determined by height:

Large Grasses — Eulalia Grass (also called Silver Banner Grass), Japanese Silver Grass, Maiden Grass, Silver Feather Grass, Zebra Grass, Porcupine Grass, Variegated Cord Grass, Plume Grass, Purple Moor Grass, Giant Reed.

Medium Grasses — Fountain Grass & Crimson Fountain Grass, Feather Reed Grass, Blue Lyme Grass, Northern Sea Oats, Tufted Hair Grass, Ribbon Grass, Switch Grass.

Short Grasses — Blue Fescue, Blue Oat Grass, Bulbous Oat Grass.

Water Garden Grasses — Golden Foxtail, Sedge, Job's Tears, Velvet Grass, Zebra Grass, Ribbon Grass.

WATER GARDENS

"The water garden has two qualities that are outstanding. It commands attention, and it creates the focal point that so many gardens lack. To anyone who has had a garden pool, no explanation of it's varied delights is necessary." — *from "Ponds and Water Gardens" by Bill Heritage.*

As gardening grows in popularity, more and more gardeners are discovering the magic of water gardening — with either an existing pond or one created with man-made materials. The plants in demand for water gardens are:

Water Lilies — the hardy varieties are perennials, and bloom every year in even the coldest climates. All are bloomers, and will bloom continuously all summer long. The tropical varieties offer a greater range of color and bloom either day or night, but require a mild climate.

Lotus — The lotus lilies are hardy perennials that have been revered as sacred plants in the Orient for thousands of years. They will tolerate colder climates, but need full sun at least six hours a day in season. The seed pods are widely used in dried flower arrangements and wreaths because of their unique appearance.

Bog Plants — These plants thrive in shallow water, usually at the edge of a pond, and add variety in height and texture to the water garden. Popular plants include Flowering Arrowhead, Sweetflag, Giant and Dwarf Rushes, Flowering Cannas and Iris, Papyrus and Cattails.

EDIBLE LANDSCAPING

All over America, lawns and formal landscaping are giving way to "edible landscaping". This new/old method integrates vegetables, herbs, fruits, nuts and ornamentals to produce healthy, home-grown food and landscaping around the home. Growers who can supply the increasing demand for edible landscaping plants should find themselves with a very profitable "niche" in the marketplace over the next few years. For more information on this fascinating topic, read "The Complete Book of Edible Landscaping", by Rosalind Creasy, or "Designing and Maintaining Your Edible Lanscape Naturally", by Robert Kourik.

ORNAMENTAL GRASSES
RECOMMENDED READING

Information Bulletin No. 64 **"Ornamental Grasses for the Home and Garden"** from Cornell University, Distribution Center C, 7 Research Park, Ithaca, NY 14850

Kurt Bluemel, well-known grower of ornamental grasses, has produced a spectacular "video dictionary" of fifty of the most popular grasses, including common and botanical names, USDA hardiness zone, planting time, best uses, and seasonal changes. It's available in his catalog, listed under "plant and seed sources."

WATER GARDENS

The Stapeley Book of Water Gardens, by Stanley Russell.
Perhaps the single most informative book on water gardening.

Ponds and Water Gardens, by Bill Heritage. *A comprehensive paperback on the topic.*

Both of these books should be available from any nursery specializing in aquatic plants.

PLANT NURSERY
RECOMMENDED READING

Bedding Plants, by Charles H. Potter, Horticultural Publishing, 111 N. Canal, No. 545, Chicago, IL 60606
This exellent beginner's manual covers all aspects of raising bedding plants. How to set up a greenhouse, simplified seeding methods, potting short course, growth regulation, management, marketing, and sixteen additional chapters cover the essentials for the new grower. Especially useful is the bedding plant chart which details sowing times, soil temperatures, seeds per ounce, days to germination, plant spacing, bloom time and hardiness for over seventy popular bedding plants.

Ornamental Plants for Profit
Growing Azaleas Commercially
U.C. System for Producing Container-Grown Plants
Agriculture Division, University of California, 6701 San Pablo Avenue, Oakland, CA 94608

These practical guides for commercial growers are typical of the information available through your state agricultural university. Call your county ag extension agent for a list of publications.

A free packet of information, including a list of specialized publications on specific plants, is available free from: **American Association of Nurserymen,** 1250 I St. N.W., Suite 500, Washington, D.C. 20005

NURSERY TRADE MAGAZINES

American Nurseryman
111 N. Canal, No. 545
Chicago, IL 60606

Pacific Coast Nurseryman
Box 1477, Glendora, CA 91740

Grower Talks
George Ball Co.
1 North River Lane, P.O. Box 532
Geneva, IL 60134

PLANT NURSERY
SEED & PLANT SOURCES

The following sources are primarily wholesale nurseries, and are not set up to deal with the general public.

Appalachian Nurseries
Box 87, Waynesboro, PA 17268
Lining out stock of unique shrubs, trees, azaleas and rhododendrons.

Maplewood Seed Company
6219 S.W. Dawn Street, Lake Oswego, OR 97034
Specialists in seeds of maples, including many rare Japanese maples.

Mellingers
2310 W. South Range Rd., North Lima, OH 44452-9731
Offers a good variety of liners and seedlings, both evergreen and deciduous, ground-covers and bulk seeds.

Mitsch Nursery
6652 S. Lone Elder Rd., Aurora, OR 97002
Specialists in lining out stock of conifers, health, azaleas, rhododendrons and other broadleaf evergreens.

Musser Forests
Box 340, Indiana, PA 15701
Tree seedlings, ground covers and shrubs, specializing in quantity orders. Retail and wholesale.

Roses of Yesterday & Today
802 Brown's Valley Road, Watsonville, CA 95076
Old, rare and unusual roses.

Verkade's Nurseries
223 Willow Ave., Pompton Lakes, NJ 07442
Nursery and lining out stock of unusual and dwarf trees.

ORNAMENTAL GRASSES
SEED AND PLANT SOURCES

Kurt Bluemel, Inc.
2740 Greene Ln., Baldwin, MD 21013 (catalog $2)

Carter Seeds
475 Mar Vista Dr., Vista, CA 92083 (free catalog)

Crownsville Nursery
P.O. Box 797, Crownsville, MD 21032 (catalog $2)

The Seed Source
Rt. 2, Box 265B, Asheville, NC 28805 (catalog $2)

Stallings Exotic Nursery
910 Encinitas Blvd., Encinitas, CA 92024 (catalog $2)

WATER GARDENS PLANT SOURCES

Lilypons Water Garden
P.O. Box 10, Lilypons, MD 21717 (catalog $4)

Santa Barbara Water Gardens
P.O. Box 4353, Santa Barbara, CA 93140 (catalog $2)

Slocum Water Gardens
1101 Cypress Gardens Blvd., Winter Haven, FL 33880 (cat. $2)

William Tricker, Inc.
7125 Tanglewood Dr., Independence, OH 44131 (catalog $2)

Van Ness Water Gardens
2460 N. Euclid, Upland, CA 91786 (catalog $2)

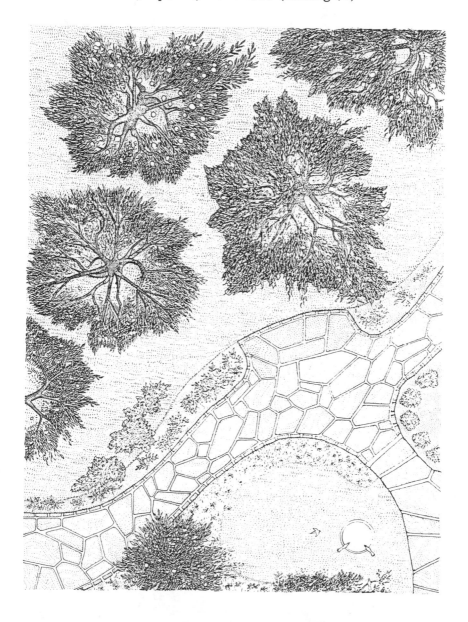

SMALL FRUITS

Great opportunities exist for part-time fruit growers, says Dr. Christopher Walsh, extension fruit specialist and professor of horticulture at the University of Maryland.

"There is a crying need for more pick-your-own blueberry, raspberry and strawberry fields near metropolitan areas," Dr. Walsh maintains. Small fruits offer a unique opportunity for a small or average size family to make a good income on small acreage.

Much of the work with small fruits can be done by hand in a family-sized operation, so less machinery investment is needed than with regular field crops. And, once the plants mature, they will usually remain productive for more than ten years. (Strawberries excepted).

In additon, small fruits currently offer a higher return than conventional field crops because of a low supply and high demand. Growers are reporting gross incomes up to $16,000 per acre per year. Specialty fruit growers in California are reporting even higher yields growing exotic fruits such as pineapple guava, mountain dorian, pummelo, carambola star fruit, Chinese dates and lychee.

One of these exotic fruits, the pineapple guava (also called feijoa) has been grown commercially in New Zealand for many years, with commercial planting just beginning in the U.S. The taste is preferred over the kiwi by most New Zealanders, and an established orchard produces 24,000 pounds per acre per year! It may be the best new fruit to plant in the next decade, if your climate is suitable. A source is listed at the end of this chapter, if you want to know more.

Only the widely adapted and grown fruits are covered in this chapter. However, there are dozens of other small fruits being grown around the country as cash crops. Consult you local extension fruit specialist for specific recommendations.

BLUEBERRIES

The next best thing to eating blueberries, says Bob Coffey, is growing, picking and cashing in on a bumper crop. This Arkansas grower has ten acres of highbush "Bluecrop" blueberries with his best acre producing over 16,000 pounds, or $16,000 worth of the plump fruits last year.

Bob also designed and built nursery propagation beds, where each year he roots over 30,000 blueberry cuttings, sold for $1.25 to $1.50 each. He has succeeded in combining small fruits and nursery production for a lucrative farming business on just ten acres.

His plants are on a four foot spacing, in rows ten feet apart, for a density of 1080 plants per acre. To insure a good market for his crop, Bob and over 150 other growers belong to the Arkansas Blueberry Growers Association. Together, they shipped over one million pounds of blueberries to markets across the country.

According to the manager of the ABGA Rick Sterne, "Most growers get started on a small scale, with pick-your-own. Then they get excited about the potential and plant more. We're seeing people realize their dreams of getting started in farming, or producing enough income to go full time."

In addition to marketing services, the ABGA also publishes a monthly newsletter, available to non-members also, for $10 per year. Write to: ABGA, P.O. Box 167, Lowell, AR 72745.

According to Dr. Neal Vincent, a blueberry specialist and chairman of the horticulture department at Delaware Valley College in Doylestown, PA, "blueberries are the BEST crop for pick-your-own. Eight to twelve thousand pounds per acre can be expected from a well-managed one acre planting."

Blueberries are easy to grow once their initial soil requirements are satisfied. The plants live and bear for up to fifty years. Acid soil is a must, with a PH of 4.2 to 5.5. If you don't have naturally acid soil, you can make it more acid with sulfur or with natural materials such as oak leaves, peat moss or well-rotted sawdust. Dig a hole two feet deep and three feet wide for each bush, and fill with a mix of the correct Ph.

Most growers start with 12" to 18" branched plants that are two to three years old. Because blueberries, due to their shallow root system, require stable soil moisture levels, a drip irrigation system insures the bushes will have ample moisture during the crucial blossom and fruit set period. It is also insurance against a prolonged summer dry spell, which could kill the plants.

After planting, mulch the new plants with 4" to 8" of any acidic material, such as straw or rotted sawdust. Mulching reduces the need for watering and helps maintain soil acidity. When the plants are two to three years old, fertilize with a balanced 10-10-10 fertilizer. Ammonium sulfate will also increase soil acidity, but organic fertilizers are preferred as they also build up organic matter in the surrounding soil.

After your plants are established, fertilize three times yearly, in March, April and May using equal amounts of organic nitrogen, phosphorus and potash. Never fertilize after May — it stimulates late shoot growth that won't be able to harden up in time for winter. In short, give your new blueberry plants lots of TLC for the first three years and they will repay you amply for the next forty.

The two blueberry types grown commercially are the "Highbush" blueberry and the "Rabbiteye" blueberry. The highbush blueberry can not tolerate the hot climate found in the southern U.S., where most rabbiteye varieties originate. Preferred varieties include: "Earliblue", "Bluetta", "Blueray", "Stanley", "Bluecrop", "Berkeley", "Herbert", "Jersey", "Dixie" and "Colville" in order of ripening.

By staggering the ripening order of the varieties, you can insure a constant on-going supply of berries, because blueberry bushes take 7-10 days to re-ripen between pickings.

Because most people pick between their knees and just above their heads, you should prune your blueberry bushes to meet these u-pick customer preferences. Top your bushes at six feet and remove brushy growth at the base of the bush.

GRAPES

The grape is a versatile fruit, with a wide variety of textures and flovors in both seedless and seeded varieties. If you plant varieties suited to your local climate, (as long as you have a 140 day frost free season) grapes can be grown almost anywhere.

The three major types of grapes are the Vinifera, American, and Hybrids. The vinifera are among the highest quality grapes for table, raisin and wine use. They are also very prone to disease. The American grapes are hardy and disease tolerant, derived from the native wild fox grape. Common varieties include "Concord", "Delaware", and "Niagara".

The hybrids are crosses between European grape varieties and native American species. They have been selected for the fruit quality of their European ancestors and the disease and insect tolerance of their American ancestors. For the best varieties, contact your extension agent or fruit specialist for suggestions.

Grapes should have a well-drained fertile sandy loam soil, high in organic matter. Grapes are adaptable, however, and will grow in a wide range of soil types. The deep roots, up to eight feet deep, make the established vines very drought-tolerant. Do not over-fertilize, as it encourages leaf growth at the expense of fruit production. A Ph of 5.5 to 6.5 is preferred.

Since your grapevines will probably outlive you, with a life expectancy of over eighty years, you should take the time to get them off to a good start! Space the new plants six to eight feet apart in a sunny spot. Remove all but the most vigorous canes when planting, pruning back to two buds. Put a five foot stake next to young vines the first year for support. A permanent trellis should be in place the second year, with stout posts on 24' centers and two to three No. 9 wires strung between them.

Because exposure to sunlight controls fruit production, you should plan your row and plant spacing to allow the maximum exposure. American varieties such as Concord will yield an average of 16,000 pounds to the acre, while wine grapes will average 10,000 to 12,000 pounds to the acre.

SELLING YOUR GRAPES

Successful marketing depends on planting the right varieties for your customers. Table grapes are sold in small ripe clusters of 1-5 pounds. To increase sales, offer recipes and/or samples of jellies and juices.

WINE GRAPES

Home winemaking is a popular hobby, as well as a good market for large purchases of grapes by the bushel or ton. Many town dwellers don't have space or time for growing, and will gladly pay a premium for fresh grapes over buying extract. The most effective selling method is a bulletin or poster at all stores that sell winemaking supplies. Consider a special co-op mailing with the store to feature up-coming harvests. Offer U-pick or delivery. Many stores like to broker wine grapes for a percentage of the gross.

PICK-YOUR-OWN

Grape growers can eliminate the cost of harvesting when the customer is willing to harvest the grapes. The customer saves money — you save time. When running a P.Y.O. for wine grapes, it is advisable to offer the use of a crusher and a press to your customers.

BONUS INCOME

To add nursery income, use your vine trimmings to propagate new grape cuttings for sale. An excellent primer on the subject is: "Propagating and Growing Grape Plants for Early Field Planting", Bulletin No. 1204, Washington State University, Cooperative Extension Service, Publication Office, Pullman, WA 99164.

KIWI

A fuzzy brown fruit with the shape and size of an egg is bringing new profits to small farms across the country. Inside the fuzzy skin is a lime-green fruit with a delicious sweet-tart flavor and ten times more vitamin C than lemons.

Native to China, where it was called Yang-Tao, the fruit was first grown commercially in New Zealand, where it was renamed the Kiwi-fruit. New Zealand now produces over eighty million pounds yearly, much of it exported. In addition to the species commercially planted in temperate areas, other tasty varieties are grown as far north as Siberia!

According to Michael Pilarski, editor of a newsletter devoted solely to the kiwi-fruit: "The Actinidias (kiwi) will most likely prove to be a fruit fast growing in popularity during the next decade. The fruits of the better selections are luscious and very high in vitamin C, Papayin (a digestive enzyme) and other vitamins and minerals. The vines are fast-growing, ornamental, and little subject to pests and disease. Those people who become involved in growing and propagating these fruits at this early stage of the game have an opportunity to economically gain plus serve the best interests of the human race.

Kiwi fruit can be processed into many food items. The fruit can be stored like apples in cold storage for two to three months, and thus the fresh marketing season is a long one. The fruits may be juiced, dried or canned. Kiwi-fruit wine made in California has been receiving awards at wine shows. The fruits can also be used in jams, jellies and preserves."

A few years ago, Lisbeth Harter retired from nursing and wanted to try a backyard cash crop. She chose the kiwi, and planted 290 vines on two acres of sandy California soil. Last year, those plants produced 50,000 pounds of fruit and grossed $28,000! Her crop was started with two year old grafted cuttings from a nursery, and started producing in another two years. The crop yield has tripled each year since, until reaching a 400 pound per plant peak.

Kiwi is capable of growing in a wide variety of soils, but doesn't like "wet feet". It is a very fast growing vine that cannot support it's own weight, and will spread up to 30 feet. In it's native forest, it uses other trees and shrubs for support. Commercial growers use a cross-arm trellis system to keep vines off the ground and support the heavy fruit loads.

Both male and female vines are needed for pollination. They are planted in rows which, if viewed from overhead, look like a tic-tac-toe board with a male plant in the center surrounded by eight female plants. They prefer a sunny spot, protected from high winds. Once established, they will produce fruit for up to 50 years. Summer irrigation is usually esential because of the shallow root system.

For those in northern climates, the "Arguta" variety of kiwi is hardy to -40 degrees. Unlike the fuzzy variety, the fruit is hairless and can be eaten skin and all, just like grapes. Growing techniques are similar, with yields lower than it's southern relatives. The Arguta variety also requires that males and females be planted to insure pollination.

RASPBERRIES

Raspberries as a cash crop have a lot of things going for them.

Quick results. A raspberry patch begins to produce the year after planting, and will bear a full crop the second or third year.

Heavy, reliable crops. Few fruits are more abundant or more dependable. Most raspberries bloom late, so a late frost doesn't usually bother them.

Little spraying. Compared with most fruits, raspberries have fewer problems with insects and diseases.

Low cost. Plants are inexpensive to purchase.

Long life. Most raspberries last 10 to 20 years or more.

There are two classes of raspberries, the reds and the blacks. The reds produce suckers from their roots to spread while the blacks develop new plants by "tipping" or bending over their long canes. Where the tips touch the ground, they take root to produce new plants. The red raspberry comes from the rose family. Although the roots live for decades, the cane or stem only lives two years. It grows the first year, flowers and fruits the second year, then dies.

The two-crop raspberries, commonly called everbearers, bear one crop of fruit in midsummer on their year-old canes, just like the regular raspberries, then produce another crop in the fall on the new canes. There is a growing interest among commercial growers in the everbearing varieties to supply fruit for the fresh markets.

Raspberries, like most vegetables, prefer lots of sun, ample organic matter, and good drainage. Prepare and clear the planting site well in advance. Your raspberries should last twenty years and preparation time is saved many times over later on. Space your rows far enough apart to allow room to till and cultivate between the rows.

Because raspberries droop so much, a minimum spacing between rows should be eight feet. For easy picking, never let your rows get wilder than two feet.

Plant new canes two feet apart in the rows, using well rotted manure. Fertilize again the following spring with more manure. At this time, you should also cut the canes back to the ground. This will produce heavy cane growth and a bumper crop the following year. Because raspberries require adequate moisture during the growing season, a trickle irrigation system may be necessary. To conserve moisture, a light sawdust mulch is helpful.

When your canes come into full production, your yields should be 6,000 to 10,000 pounds per acre for red raspberries and somewhat less for the black raspberries.

Many commercial growers have a signed contract with a processor before each season. Fresh marketing is also gaining in importance with recent advances in packaging and shipping. For example, you can now buy fresh American raspberries flown in daily in season, in Paris and Tokyo. Don't ask how much.

Additional fresh markets include retail stores, wholesalers, nursing homes, schools, hospitals, restaurants and food co-ops. And of course roadside stands, pick-your-own and farmer's markets.

STRAWBERRIES

Strawberries are one of the best small fruits for a backyard cash crop. Because of the popularity of strawberries, you should easily sell all you can grow, either harvested or pick-your-own. Also, costs of establishing a strawberry patch are low, compared to the other small fruits.

Varieties are very important, and time and effort should be taken to select varieties that meet your requirements and your customer's needs. When planning a pick-your-own planting, consider early maturing varieties, possibly used with plastic soil mulches or row covers to provide **early** berries to attract customers and stretch the P.Y.O. season.

A good place to start in selecting varieties is to visit other P.Y.O. growers and find out what varieties are successful with them. To spread out the picking season, plant early, mid, and late season varieties. Other qualities to look for are a high yield, and a berry that is a good keeper on the stem and in the basket. In addition, a good P.Y.O. strawberry should ripen evenly, with no green tips, not rot in wet weather and pick easy. (Customers don't like to tug on a berry.)

Plant scientists at the U.S. Department of Agriculture's research center have developed two promising new "day-neutral" strawberries which fruit in the spring like other varieties, then just keep on bearing once they've started. Called Tristar and Tribute, they are relatively disease resistant and good producers.

While you are checking out new varieties to plant, you will need to prepare your planting site. First, go to your extension service and have your soil tested. Strawberries do best with a soil Ph of about 6.5. The soil test will also tell you the amount of phosphorus and potash required. Strawberries require ample nutrition to produce well.

For a pick-your-own strawberry field, rows are usually spaced four feet apart. If you plant closer, the runners tend to fill in the walkways between the rows, and a lot of berries get stepped on by pickers. Set your plants eighteen inches apart in the rows. For every plant you set out, you should get two quarts of berries. An old grower's rule of thumb is to plant when the leaves on the trees begin to show some green.

The yield per acre from a well-managed strawberry field will range from 6,000 pounds to 16,000 pounds. According to university fruit specialists, the average yield for P.Y.O. fields is 10,000 pounds per acre.

SMALL FRUIT MARKETING

Many people will drive miles out of their way to find fresh picked farm-direct produce, especially fruit. In our area, the county extension service publishes a map each year showing the roadside farmer's markets and pick-you-own farms throughout the six-county area. It's become a popular and money-saving recreation for city and suburban families to pile in the station wagon to go picking or produce shopping where crops are fresh picked and prices are less than retail.

TIPS FOR A SUCCESSFUL P.Y.O

Plan your field with the customers in mind, with easy access, adequate parking and information signs.

Enclose your field with a fence when possible, so customers must use a check-in/check-out area.

Supervise pickers. Assign rows and check to make sure all the ripe berries are being picked.

Keep your customers supplied with empty picking boxes. Pick up full boxes and tag with customer's name. You want your customers to concentrate on picking berries, not having to worry whether they can carry all their full boxes!

Make your customers as comfortable as possible so they will stay and pick longer. This means providing restroom facilities and drinking water.

Sell by the box, not weight, for quick and easy check-out. This way you simply count the number of boxes and multiply by the price per box.

Signs are very important — they are the first contact you have with your customers, and first impressions are lasting. Try to have small signs along the way. For example: a strawberry and an arrow, with the distance to your field.

Use re-usable containers printed with your name and phone number. Offer credit for each returned basket.

PICK-YOUR-OWN ADVERTISING

For every "pay-n-go" customer, you will have three that want to "pick-their-own." This area of direct farm sales is growing by leaps and bounds because customers can buy "wholesale", and you can save the cost of harvesting. Here are some pointers aimed at getting those P.Y.O. customers to your fields in the first place.

FREE ADVERTISING

You can get free publicity if you let the media people know what you are doing. Especially in small towns, specialty crops are almost always newsworthy.

MAILING LISTS

Have your customers fill out a card with their name, address, phone number and the crops they are interested in. Emphasize that "preferred" customers will have first pickings; this encourages sign-ups. Send out postcard mailers to your mailing list a week or two before the crop is ready. This builds loyalty and repeat business.

DISPLAY ADS

A modestly priced ad in the local weekly or shopper lets everyone know what you are up to. Use a large headline to get attention; for example **"U-PICK BLUEBERRIES"**. Another attention-getter is a hand lettered ad. It will stand out, and the newspaper will not make copy mistakes. What yu printt iz whatt u git!

CLASSIFIED ADS

Market research has established that folks are more inclined to read the classified ads in local papers than display ads. You should run a continuous classified through your harvest season. For example: **U-pick jumbo strawberries until July 10 at Valley Farms, 1800 Valley Highway. Call 555-1201 for current picking information.** Establishing a cut-off date in your ad helps overcome procrastination. Providing a phone number lets you tell folks if you're sold out.

LOCAL RADIO SPOTS

Away from larger metropolitan areas, the cost of radio "spot" advertising can be quite reasonable. The advantage is your ability to update crop information easily as conditions change, not always possible with a weekly newspaper. Talk to your local station and set up a program. Strawberries, for example, with a short harvest season of less than a month, could use "Berry Reports" four to six times per day for the entire harvest season.

NURSERY PRODUCTS

Many growers supplement their growing income during the off-season by propagating the cuttings and trimmings from their crops. Bob Coffey, the blueberry grower, is a good example of what is possible, with his gross income from spare-time propagation approaching $40,000 per year.

New varieties are easiest to sell, as everyone wants to try at least one. Keep in touch with your extension agent to find out what's new and different. And remember that the prices you get from your plants can sometimes double each year as they grow and mature. A local grower specializes in mature four year old blueberry bushes and charges $10 each, when you can get a rooted cutting for just $1.50!

Other growers have focused on hard-to-find varieties, and sold their plants by mail-order. Unless you have a unique, scarce, or patented variety, you are better off selling your nursery stock locally. Mention in your harvest ad that customers can also purchase plants to take home.

VALUE ADDED PRODUCTS

A local grower, Wax Orchards, got fed up with low prices from the juice processors and decided to start manufacturing their own products. Now, they must buy from other growers to meet the demand for quality organic juices, jams, jellies and fruit leathers.

Test-market your products before you invest a lot of time and money. Put up a batch of strawberry preserves or raspberry-apple juice or kiwi-jelly and get it on the shelf at your local grocery, food co-op or gift shop. Don't scrimp on the packaging! Find a capable commercial artist to do your labels — it could make the difference between success and failure.

PRODUCER CO-OPS

Many crops have grower/producer marketing co-ops or associations. In addition to finding, enhancing and enlarging crop markets, they can also help with grading, packaging and extended storage. Check in your area to see if a local marketing group is active. For those who prefer to concentrate on growing rather than selling, a co-op can be ideal.

RECOMMENDED READING

Current information on growing and marketing small fruits is available from:

USDA Fruit Laboratory
Agricultural Research Center West
Building 005, Room 305
Beltsville, MD 20705

In addition, the USDA has a staff of research specialists avaiable, free of charge, to get you answers to technical questions on small fruits or other growing topics. If they don't have the answer, they will refer you to another expert who can help. Contact:
USDA Information Office, Room 230E,
Washington, D.C. 20250

Your local extension agent will have a list of regional publications available on small fruits. They also can refer you to the extension fruit specialist (every state has several). This is a good starting place for information on the crops that interest you.

Each year, the University of Illinois holds a conference for growers called the Illinois Small Fruit School. If you can't attend, order the printed transcript for $3. Write to: Dr. J.W. Courter, Dixon Springs Center, Simpson, IL 62985.

The two complete texts on blueberry production are "Blueberry Culture", edited by Paul and Norman Childers, horticulturists at Rutgers University; and "Blueberry Science" by Paul Eck. Write: Rutgers University Press, Distribution Center, P.O. Box 4869, Hampden Station, Baltimore, MD 21211

Kiwifruit Culture, by P.R. Sale, is a complete reference guide to kiwi, with cultural requirements, pruning and trellising, harvest and packing, disease and pest control. The only all-inclusive book currently available on Kiwi-fruit. Available through: AG-ACCESS, P.O. Box 2008, Davis, CA 95617.

Kiwi Growers of California, at P.O. Box 922, Gridley, CA 95948, publishes a newsletter for commercial growers, devoted to production and marketing.

Actinidia Enthusiasts Newsletter, at P.O. Box 1466, Chelan, WA 98816, also publishes a newsletter for kiwi-fruit growers nation-wide. A sample issue is $3.

SMALL FRUITS SEED & PLANT SOURCES

BLUEBERRIES

DeGrandchamps Blueberry Farm
15575 77th St., South Haven, MI 49090

Blue Star Laboratories
Route 13, P.O. Box 173, Williamstown, NY 13493

Hartman's Plantation
310 60th Street, Grand Junction, MI 49056

Rayner Brothers
P.O. Box 1617, Salisbury, MD 21801

GRAPES

Burntridge Nursery
432 Burnt Ridge Road, Onalaska, WA 98570

Country Heritage Nursery
P.O. Box 536, Hartford, MI 49057

Raintree Nursery
391 Butts Rd., Morton, VA 98356

KIWI-FRUIT

Stanley & Sons Nursery
11740 S.E. Orient Dr., Boring, OR 97009
Largest collection of Kiwi varieties in U.S.

Edible Landscaping
P.O. Box 77, Afton, VA 22920

Friends of the Trees
P.O. Box 1455, Chelan, WA 98816

Northwoods Nursery
28696 S. Cramer Rd., Molalla, OR 97068

RASPBERRIES

Brittingham Plant Farms
P.O. Box 2538, Salisbury, MD 21801

Dean Foster Nurseries
P.O. Box 127, Hartford, MI 49057

Malieski Berry Farms
7130 Platt Road, Ypsilanti, MI 48197

STRAWBERRIES

Ahrens Strawberry Nursery
R.R. 1, Huntingburg, IN 47542

Brittingham Plant Farms
P.O. Box 2538, Salisbury, MD 21801

W.F. Allen Co.
15 Strawberry Ln., P.O. Box 1577, Salisbury, MD 21801

Country Heritage Nursery
P.O. Box 536, Hartford, MI 49057

Rayner Brothers
P.O. Box 1617, Salisbury, MD 21801

N.Y. State Fruit Testing Co-Op Assoc.
Geneva, NY 14456

PINEAPPLE GUAVA (FEIJOA)

Callender Nursery Co.
P.O. Box 2326, Chico, CA 95927
In addition to supplying plants to commercial growers, they publish an informative booklet on this new crop.

Hartman's Plantation
310 60th St., Grand Junction, MI 49056

SPECIALTY CROPS

BAMBOO

For thousands of years, bamboo has been an everyday part of Asian culture, providing food, shelter and raw materials for a wide variety of products. In Japan and China, "Moso" bamboo is cultivated for it's edible shoots, producing a continual harvest of three to ten tons per acre.

Many bamboo species are used in construction projects because of their incredible strength, approaching that of steel. It's not uncommon in the Orient to see a ten story bamboo scaffolding surrounding a new highrise apartment building. Other common uses include garden stakes, fence and orchard posts, irrigation tubing, gutters, flutes, furniture, and concrete and plaster reinforcing.

Bamboo is now being "re-discovered", and most established suppliers cannot keep up with the demand. The supplier with the greatest diversity of plants, Hermine Stover, of **Endangered Species,** says in her current catalog: "We must have done something right because our volume has multiplied into the unendurable category, despite the fact that we hardly advertise, and do nothing to encourage the growth of our mailing list. Ours is still a hand-made low-overhead cottage industry."

In her catalog, Hermine lists the "Top ten most popular bamboo in America", measured by her mail-order sales.

1. Phyllostachys Aureosulcata — Yellow Grove Bamboo.
2. Phyllostachys Nigra — Black Bamboo.
3. Bambusa Oldhami — Giant Clumping Timber Bamboo.
4. Phyllostachys Vivax — Smoothsheath Bamboo.
5. Bambusa Glaucescens "Alphonse Karr".
6. Sasa Palmata — Palmate Bamboo.
7. Bambusa Vulgaris Vitata — Giant Striped Bamboo.
8. Sasaella Glabra — Dwarf Clumping Bamboo.
9. Phyllostachys Aurea-Golden Bamboo.
10. Phyllostachys Bambusoides — Giant Timber Bamboo.

Bamboo is tolerant of a wide variety of soils, but prefers a slightly acid well-drained soil. Some varieties, such as P. Vivax and P.

Aureosulcata prefer full sun. Others, such as P. Nigra, prefer partial shade. Most of the larger varieties of bamboo produce edible shoots, which are harvested in the spring before they grow beyond 12" high. The shoots are boiled first, then eaten fresh or salted or pickled. The current wholesale price at the produce markets on the West Coast is $2 per pound, fresh.

Because bamboo is among the most beautiful of all plants, it is a popular year-round privacy screen with it's dense growth.

When mature, bamboo cane can be sold by the foot, just like lumber, for arts and crfts projects, fences, trellises and patio furniture. U.S. Growers cannot currently keep up with the demand, so most of the cane is imported. A current price for a small 3" x 12' cane is $24!

BONSAI

From tiny trees the size of your thumb to mature maples just eighteen inches high, the world of bonsai contains many surprises. Bonsai is the Japanese technique of stunting trees and shrubs without altering their natural appearance.

Centuries ago, the Japanese started collecting the gnarled and twisted trees and shrubs that had been naturally dwarfed by a harsh existence in the rocky crevices of the mountain or seashore cliffs. A growing scarcity of specimens in the wild encouraged growers to develop methods of training that duplicated the wild specimens, and the art of bonsai was born.

Most bonsai trees and shrubs are developed from regular nursery stock while still young enough or small enough to train. Some of the most popular are:

EVERGREENS	**DECIDUOUS**
Atlas Cedar	Trident Maple
Hinoki Cypress	Japanese Maple
Dwarf Norway Spruce	Japanese Hornbeam
Dwarf White Spruce	Chinese Hackberry
Yeddo Spruce	Maidenhair Tree (Ginkgo)
Japanese White Pine	Deciduous Japanese Holly
Sargent Juniper	Pieris Japonica
Bar Harbor Juniper	Wisteria Floribunda
Prostrate Juniper	Sargent Crabapple
Dwarf Japanese Yew	Scarlet Firethorn (Pyracantha)
	Paper Birch

FOUR WAYS TO GROW BONSAI

If you are planning to grow plants to sell to bonsai hobbyists, there are four ways to specialize:

STARTER TRAYS

Many growers sell trays of starter plants; two or three dozen per tray. Many bonsai hobbyists prefer to grow and train their specimens from this seedling stage, and will purchase in quantity. Give purchasers a choice by offering trays of all one kind or assorted for 25 percent more. Most of your tray customers will be retailers such as garden centers, florists and bonsai clubs.

STARTER PLANTS

The same starter plants, a year or two older, and in 2 1/4" to 4 1/2" pots, are sold individually or in small assortments of six or so to bonsai fanciers who want to train their own plants. Prices vary according to age, shape of the plant and scarcity, but can be as much as $40 per plant.

TRAINED PLANTS

Bonsai plants that have been "trained" to their first ceramic pot are in wide demand, as the buyer can see what the plant will look like as a bonsai subject; yet it will be young and small enough to still be affordable. Prices can range from $10 to $60, depending on size and beauty.

Also popular in this price range are "mini-gardens" of two to five plants artistically arranged in one bowl. Many florists who purchase your starter trays will use these plants to create their own mini-gardens for resale.

SPECIMEN PLANTS

As your starts mature and you train them, set aside a few choice plants to mature for future sale as specimen plants. Prices are determined mostly by artistic appeal at this stage, but most good specimens fetch hundreds of dollars.

If you choose to become involved with bonsai, think about selling authentic bonsai pots as a sideline. You can work with established importers or, better yet, find a local potter to make your pots to bonsai specifications. Keep in mind the current markup on bonsai pots is as much as 500 percent.

DRIED PLANTS

For centuries, dried plants have been used for decoration and fragrance. Garlands and wreaths of blue delphiniums and lotus blossoms, still in good condition, were discovered by archeologists in ancient Egyptian tombs.

Our ancestors used dried flowers to conceal household odors and brighten up the house in the winter months. Since then new plants have been discovered, and most important, new methods of drying have been developed to preserve even the most fragile blooms.

Because of the popularity of dried floral arrangements, the commercial market for dried flowers is large and growing rapidly. Baby's Breath, for example, is a great bouquet extender. It's been called the "hamburger helper" of the floral industry. When you care enough to send roses, but not enough to send a dozen, most florists will attempt to make you look good by mixing the roses with a handful of Baby's Breath.

Betsy Williams, a grower of dried flowers, has to buy additional plants from other growers because, as she says: "I can't grow enough statice or baby's breath for my own needs!"

The two most common methods used to preserve your harvest are air drying and silica gel. In the air drying method, strip foliage from the fresh cut flowers. Tie loosely in small bundles and hang upside down in a shaded, well-ventilated area. Harvest your flowers on a dry sunny day after the dew has evaporated from the blossom. Using an inexpensive box fan to increase air circulation will reduce drying time greatly.

Hard-to-dry flowers can be preserved with a special drying agent called silica gel. It will dry the difficult flowers quickly, preserving both form and color. The silica gel acts as a "dessicant", which draws moisture out of the plant. For "instant" drying, silica gel can be combined with a conventional oven or microwave.

WHAT TO GROW

Listed below are many of the flowers, grasses and seed pods in demand today, together with harvesting tips. To test for dryness, check to see if the petals are crisp to the touch.

Acroclinium (Everlasting Daisies) This handsome small flower, about 1" in diameter, is used in bouquets and is known for holding it's color indefinitely. Hang to dry for about one week.

Baby's Breath (Gypsophilia) Easy to dry and very popular, this should be picked at full bloom and hung for almost ten days to dry. Since it's a perennial, you will not need to replant once the first planting is established.

Celosia (Cockscomb) This lovely flower comes in both created and plumed varieties in a rainbow of colors. Hang to dry with short stems to speed drying; about six stems to a bunch. Harvest just before seeds are ripe. After hanging two to three weeks, store in a box or bag to prevent color fade.

Globe Amaranth (Gomphrena) Also called "clover flower", because the blooms resemble clover. Available in a variety of soft pastel colors. Harvest with stem when the flower is mature. Strip all but the top leaves close to flower. Tie in bunches of 10-12 and hang for about 2-3 weeks to dry.

Globe Thistle (Echinops) A striking everlasting with globe shaped violet blue flower heads. Harvest when the heads start to show blue, but before the flowers are fully open. Leave about a foot of stem, strip the leaves, and hang to dry for about two weeks.

In addition to these widely used flowers, there are many other lesser-known but very salable crops to consider growing. Here are just a few:

Ammobium — A very showy everlasting.

Anaphalis — Small silver-white flowers with yellow center. Ideal for drying.

Armeria — Also called **Thrift,** this low-growing flower has an abundance of golfball-sized blooms.

Chinese Lanterns — Also known as **Physalis,** it produces a bright red hollow fruit much in demand for winter bouquets.

Corn — Popular varieties include **Calico Corn, Rainbow Corn, Miniature Indian Corn and Red Strawberry Popcorn.** The brightly colored kernels of orange, purple, red, blue, yellow and black add a decorative touch to arrangements.

Gomphocarpus — Colorful bronze to green-yellow fruit pods.

Nigella — Or **Love-in-a-Mist** produces a multitude of tiny flowers in a web of greenery.

Sea Holly — These unique pineapple-shaped blue flowers are easy to hang and dry.

Wood Rose — The dried flower looks like a rose carved out of wood, with a polished satin-brown color.

Yarrow — Long stemmed golden saucer-shaped heads; a staple in many arrangements and lasts forever.

Ornamental Grasses are popular in arrangements, and easy to grow and dry. Just pick when mature and hang for about a week to dry. Common Varieties include:

> **Animated Oats** — used in large bouquets.
> **Cloud Grass** — wispy and delicate.
> **Feather Grass** — mixed with flowers.
> **Foxtail** — reddish gold cylinders.
> **Job's Tears** — pearl grey seeds.
> **Pearl Grass** — purple spikes/white beard.
> **Pampas Grass** — large distinct plumes.
> **Quaking Grass** — a nodding grass.
> **Squirrel Tail Grass** — 3" feathery heads.

Lunaria (Honesty and Silver-Dollar) Harvest this distinctive plant with a long stem after the discs have turned tan and ripened. Hang until shells are dry, then peel off the outer layer to expose the silver petals. Handle with care, as they are fragile and damage easily.

Statice — Available in a rainbow of colors and quite popular with the florists. Harvest with a one foot stem, stripping the lower leaves, and hang for at least two weeks.

Strawflower (Helichrysum) Easy to dry — just harvest before the flower is fully open, strip the leaves from stem and hang to dry for two or more weeks.

Xeranthemum (Immortelle) White, lavender and pink papery flowers. Harvest at full bloom, hanging for about a week to dry.

MARKETING DRIED PLANTS

The primary markets for your dried flowers and plants will be to the florist trade. If you have ample acreage, you may want to sell your crop to the wholesale florists or to the export market.

Growers with limited growing space should concentrate on variety in order to supply local outlets such as gift shops, florists, garden centers, nurseries and variety stores.

A much more profitable "value-added" approach would be to use your crops to make bouquets and arrangements. The best markets are gift shops, variety stores, supermarkets, garden centers and florists. Find out what's selling, how much it costs and how it's packaged before you plant your crop.

When selling your crops or products, get cash! Remember, you're a grower, not a banker. If the buyer is hesitant to pay on delivery, offer to consign a trial dozen or two. Follow through in two weeks to check sales.

When you wholesale your products, the standard discount to retailers is 40 percent. For example, a bunch of dried flowers that retail for $10 would be sold wholesale for $6. When you consign your stock, the retailer has no cash tied up in stock, so the standard discount is less, usually 25 to 30 percent.

Cheri Woodard, the owner of Faith Mountain Herbs, says her best sellers are wreaths. Says Cheri, "We are constantly upgrading our merchandise and the prices go up correspondingly, but people seem willing to pay the difference. Last year, our wreaths were $30, they are now $60.

Remember that anywhere holiday decorations are sold is a potential market. Christmas wreaths can produce an excellent income during the "off" season.

If you decide to sell your crop to florists, write to: **Teleflora, 12233 W. Olympic Blvd., Los Angeles, CA 90064.** They can supply a membership directory (currently over 17,000 members) and other services to growers. Subscribing to their monthly magazine for florists **"Flowers And"** will help you keep up with trends in the floral industry.

To check on current export markets for your dried flowers and plants, contact the U.S. Department of Agriculture, Foreign Agricultural Service, Room 5071 South Building, Washington D.C. 20250. Their job is to expand export sales, and they can help you research the demand overseas for a crop, then help you market the crop when you've harvested it. All at no cost to you!

SPECIALTY MUSHROOMS

While largely ignored in North America, for centuries the Japanese have been using the world's largest renewable food resource — plant cellulose. This woody material that acts as the support system for plants requires the proper digestive system to turn it into food, found only in fungi and insects.

With fungus, dead trees, sawdust, wood shavings and straw can be transformed into delicious fresh mushrooms. This proven method was first used by ants and termites to grow edible mushrooms on woody debris. Thousands of years ago, humans copied the ants by growing mushrooms on rice straw in Asia. Since then, Asians have domesticated several mushrooms, growing them on straw, cordwood and sawdust.

Today, over 200,000 Japanese growers produce the **Shiitake** mushroom on small family farms. The crop is harvested in the spring and fall and most is dried for year-round domestic use and for export sales. (Over one billion dollars per year worth!) The Shiitake is prized not only for it's meaty taste and rich flavor, but for it's medicinal properties, as it contains several proven anti-tumor compounds.

Reseachers at the U.S. Forest Products Laboratory call the Shiitake mushroom "a promising new industry for the United States." It's an ideal crop because it uses only woodlot thinnings, doesn't require fertilizers or pesticides, has a low start-up cost, and is adaptable to most parts of the country. According to the U.S. Forest Products Lab, "The climate for outdoor cultivation is acceptable in all areas with adequate rainfall."

The outdoor growing method is quite simple. Logs three to four feet long and three to six inches in diameter are cut, ideally in winter when the sugar content is highest.

Then the logs are innoculated with mushroom spawn, usually available as solid wood plugs. Holes are drilled for the plugs along and around the log. Then the plugs are inserted and sealed with wax to prevent contamination and drying.

The innoculated logs are stacked in a shady spot with good air circulation. After eighteen to twenty four months, the logs will begin to fruit. According to Bob Harris, author of "Shiitake Gardening", each log will produce about two pounds of mushrooms over it's fruiting life.

At a fresh price of $4 per pound, this means a return of about $2,000 per cord of logs over the life of the logs. According to Daniel Kuo of the Mushroom Technology Co., a medium sized shiitake farm produces about five thousand pounds of mushrooms annually.

To reduce the two-year waiting period for outdoor production to less than five months, researchers at the University of Toronto's botany department have published a formula for indoor shiitake growing.

Mix a growing medium of 80 percent hardwood sawdust, 17 percent wheat bran, 1 percent molasses, 1 1/2 percent yeast extract, 1/4 percent limestone and sufficient water for a 60 percent moisture content by weight.

This growing medium is put in log shaped plastic bags and kept at 75 degrees for four months. The the "logs" are submerged in 60 degree water for 24 hours and then kept at 50 to 59 degrees. Mushrooms will begin to grow out the open end of the plastic bag within two weeks.

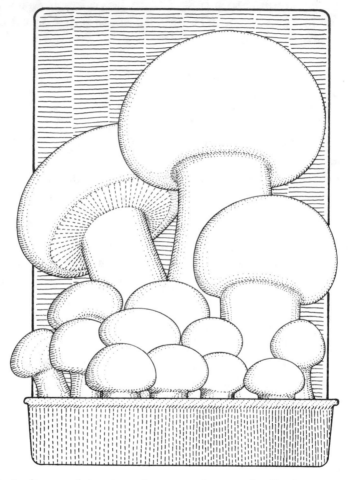

With indoor cultivation, logs can be artifically forced to fruit three or four times a year. This allows a grower to produce a crop when prices are highest, rather than as the weather permits.

OYSTER MUSHROOMS

Oyster mushrooms are another variety of specialty mushrooms, long popular in Europe, and that is gaining in appeal in North America. Oyster mushrooms are also fast growing and produce heavy yields. Fifty pounds of wheat straw, for example, will produce almost fifty pounds of mushrooms. One California hobby grower produces over three thousand pounds a year (at $4 a pound) in a less than 100 square foot shed!

Dr. Robert Todd, head of the Microbiology Department at South Dakota State University (Brookings, South Dakota 57007) has been working on an oyster mushroom production and marketing program since 1985.

According to Todd, oyster mushrooms have several advantages over common button mushrooms, including: the growing medium or substrate can use waste products readily available on a farm such as wheat straw or corn stalks; the substrate can be used as a livestock feed after it has been used to grow a crop of mushrooms; and most important, oyster mushrooms can be marketed at a price three to five times that of common mushrooms.

Dr. Todd has been test marketing the oyster mushrooms in eight and sixteen ounce shrink-wrapped trays, wholesaling for $2 and $4, with a retail price of $6 per pound. He estimates a grower should be able to produce oyster mushrooms for 50 cents to $1 per pound. To learn more about this program, write him at the address given above.

We talked to several commercial growers who are excited about the prospects for this new crop, including one in Canada who currently produces six thousand pounds a week and reports more orders than they can fill.

John Meek, of Merigold, Miss., picked oyster mushrooms to grow for market because they provide a quicker return than other gourmet mushrooms, fruiting in just a month. John and his wife converted an unused barn to mushroom production, insulating the walls to control temperatures to 50 to 60 degrees. They use commercial spawn to innoculate bermuda grass that has been pulverized, then pasturized in 150 degree water for three hours. Then the grass is compressed in a household trash compactor into 40 pound bales. The innoculated bales are wrapped in plastic and stored in the barn for three weeks. Then the plastic is removed and the bales start producing 5 to 6 pounds of mushrooms a week for six to eight weeks. The oyster mushrooms are sold to restaurants in the area and to other local customers. John says, "Once people eat oyster mushrooms, they keep coming back for more."

MARKETING SPECIALTY MUSHROOMS

Your best prices will come from local buyers, such as restaurants, grocers, food co-ops and health food stores. Get together with other growers in your area and start a co-op marketing association, if there isn't one now. This will give you and your fellow growers leverage when dealing with retailers and wholesale produce buyers. It also allows you to focus on growing, rather than selling.

If your location is suitable, consider a "pick-your-own" operation. Folks are fascinated by mushrooms, and you should get an abundance of regular customers from your initial advertising. Also consider selling at your local farmer's market.

DEMONSTRATIONS

Arrange with local grocers and supermarkets for a demonstration to establish or increase demand for your exotic mushrooms. Spend a morning feeding your tasty mushrooms to customers in the store. Wear a white apron, and bring a small stand, an electric skillet, butter, herbs, toothpicks, napkins and lots of mushrooms! A flyer to pass out with recipes and general information on your mushrooms is important too.

DRIED MUSHROOMS

There is a year-round market for dried shiitake mushrooms in the $16-$24 per pound range. This allows you to sell your best mushrooms fresh for a premium price, then dry your lower quality harvest to sell later in the season, or when the prices suit you. Because dried mushrooms will keep for a long time when dried and stored properly, you have flexibility on when and how you sell your crop.

SEEDS

Back in 1975, a neighbor gave Bill and Marylou Schmidt a handful of turnip seeds to try in their garden. The unique turnip dated back to a turn-of-the-century farmer who brought the seeds from his native England. The sweet flavor and creamy color of this turnip caught their fancy and they put up 150 packets of seed to sell in addition to their other cash crop, Christmas trees.

The seeds sold out in ten days. "We didn't expect that," Marylou commented "but we wanted to follow it up." So the next year they planted extra rows, sold 3500 packets of seed — all they had — and had to return hundreds of orders!

Bill and Marylou Schmidt's half-acre specialty seed crop is typical of many small growers across the country. In Idaho, dozens of small growers are successfully growing flower seed under contract to the major seed companies. Because of the attention required, most growers have just three or four acres in production, growing cosmos, California poppies, asters, zinnias, calendula, bachelor buttons, dahlias, marigold and nasturtium flowers for seed.

Across the country, hundreds of small growers are planting and harvesting regional wildflower mixes, which arc enjoying a revival in popularity with the trend towards natural and no-mow landscaping. Mixes are grown under contract to seed companies or sold direct to landscapers, developers and homeowners for low-maintainence landscaping projects. Most mixes retail for $25 to $60 per pound.

During the last century, seed production was a local or regional activity. Over the last twenty years, large "agri-business" seed companies have been systematically reducing the varieties available, pushing hybrids and patented plant stock to create a near-monopoly in many varieties.

This has created a seed crisis, and at the same time, an opportunity for the small grower to preserve and distribute endangered varieties. For the sake of a healthy, sound and diverse agriculture, regional seeds that can do well in your "micro-climate" are essential.

To learn more, write to: Seed Savers Exchange C/O Kent Whealy, 203 Rural Ave., Decorah, IA 52101. The exchange is a non-profit organization of gardeners dedicated to finding and spreading heirloom vegetable varieties before they are lost.

They publish a newletter each year which contains the names and addresses of it's hundreds of members and the old or foreign or unusual vegetable varieties each has to offer. Members trade seed for postage, while non-members send a dollar and a self-addressed stamped envelope with requests.

To obtain the latest copy of the "True Seed Exchange", including seed saving guide, companion planting guide, and membership information, send a $2 donation to the above address.

SPROUTS

For those who want to raise a backyard cash crop, but have limited space, sprout growing is an ideal business. According to Monica Hadley, of Sunsprout Systems, this unique business can be started in a corner of your basement or garage with as little as $2500, gross up to $4500 per month, and provide a net income of up to $2200 monthly to the grower. Here are some tips from Monica for the prospective sprout farmer.

Sprout growing is an ideal business for many people. The product is beneficial to consumers, and every store, restaurant and food-service is a potential customer. Sprouting is equally successful as a sideline business.

Ten years ago, pioneer sprout growers had their sales efforts greeted by comments like "Alfalfa — isn't that what horses eat?" and "How many acres do you grow?" Today, fresh sprouts are commonly recognized as a tasty and nutritious addition to salads and sandwiches. During the intervening decade, sprouts have been recognized by many as a unique business opportunity.

MARKETING SPROUTS

Strongly consider direct sales to stores, restaurants, cafeterias and institutions such as hospitals if any of the following indicators are present:

1. Major wholesalers for your area are located in another city.

2. Large wholesalers move a relatively small volume (less than one case per store per week) of sprouted products. The wholesaler carries sprouts as a convenience for those customers who request them but make little effort to sell to others.

3. A base of potential customers sufficient to meet your sales goals exists within driving distance. The major problems with direct delivery are higher delivery costs and more time required to serve more customers, at a smaller return per customer. Higher prices can make up for the higher costs, and the time can be turned to advantage. Every delivery stop is an opportunity to educate the retailer or cook, to ensure the quality and freshness of your product on the shelf and to increase the volume of sprouts you sell.

"If you are considering establishing a sprouting business, the first step in assessing your potential customer base is a telephone survey to get a general picture of current market volume, prices, sources and quality. Cities of 100,000 or more are likely to have produce wholesalers which service independent stores and restaurants. Contact restaurants and cafeterias which feature salad bars or gourmet sandwiches.

Next, go on a buying trip to stores in your area, making notes of price, quality and condition. Talk to the produce manager and find out what he thinks of the sprouts he carries, how much he sells, and if there is interest in other varieties of sprouts.

Is the current price at a profitable level? A common retail case size for alfalfa sprouts is one dozen 4 oz. packages, or three pounds per case. Wholesale prices range from $2.50 to over $6.00 per case. Alfalfa sprouts cost only about 20¢ per pound in direct costs (seeds,utilities and labor) to produce, plus between 15-50¢ per pound to package. You can see how important (and costly) your choice of packaging can be!

Keep in mind — price is only one factor in a buyer's decision. If prevailing prices are low, you can compete at a higher price by emphasizing service, product quality, packaging or marketing.

Sprout growing is farming. With equipment such as that produced by manufacturers like Sunsprout Systems, the sprouts grow themselves, requiring little attention during the growth cycle. Seeding, sanitizing, packaging and delivery, however, are labor intensive and repetitive. You should enjoy growing things and experimenting to get optimum results. The most successful of growers remain involved, even if they have employees for the physical labor. An experienced mid-sized grower will average 20-25 pounds of sprouts per hour."

SPECIALTY CROPS
RECOMMENDED READING

BAMBOO

The Book of Bamboo by David Farrelly
This lovely and inspiring book is a good introduction to bamboo. It's available from: Endangered Species, P.O. Box 1830, Tustin, CA 92681-1830, or can be ordered through any good bookstore.

BONSAI

A series of basic bonsai handbooks is available through the Brooklyn Botanic Gardens. Included are:

Dwarfed Potted Trees — The Bonsai of Japan

Bonsai — Special Techniques

Bonsai for Indoors

Japanese Gardens and Miniature Landscapes

Each handbook is about $3; get a free copy of their current publication/price list. Brooklyn Botanic Garden, 1000 Washington Avenue, Brooklyn, NY 11225

DRIED FLOWERS

The two best books on preserving and arranging dried flowers and plants currently available are by Roberta Moffitt, a national authority on flower preservation. Both books use easy-to-understand instructions and pictures for virtually foolproof results. The author also publishes a free newsletter/catalog.

Step-by-Step Book of Preserved Flowers
Step-by-Step Book of Dried Bouquets

Available from: Roberta Moffitt, P.O. Box 3597, Wilmington, DE 19807

MUSHROOMS

Growing Shiitake Commercially by Bob Harris

Illustrated procedures for cultivation of shiitake mushrooms on logs using dowel plug spawn. Using this method, the time for spawn run is shortened from the usual two years to just four months. This 72 page booklet is an excellent introduction to the subject. From: Mushroompeople, P.O. Box 159, Inverness, CA 94937

Bob Harris has just completed a new video tape about cultivation of Shiitake in Japan. The tape is an excellent introduction to the subject, and contains very important growing tips from his many trips to Japan to study the expert growers there.

Is Shiitake Farming for You?
A basic manual for starting your own farm covering step-by-step production methods, economics and marketing. From Far West Fungi, P.O. Box 428, South San Francisco, CA 94083

Shiitake News
Published by the Forest Resources Center, at Route 2, Box 156-A, Lanesboro, MN 55949; this is the only newsletter currently available about specialty mushrooms. The center director, Joe Deden, covers all the current information available on the topic, including growing and marketing tips, sources of supply, current research and much more. In addition, the center offers day-long seminars on growing and selling, and field trips to Japan to study cultivation techniques. They have also just published their "Shiitake Mushroom Marketing Guide." The guidebook contains complete step-by-step information and worksheets, in an easy to use format. Write for a current price.

SEEDS

Seed Production
This 700 page book, by Hebblethwaite, is the "bible" of seed production. It contains the most recent information on all types of seed production by the top researchers. Highly recommended for anyone serious about the subject. Available thru: AG ACCESS, P.O. Box 2008, Davis, CA 95617

Growing Garden Seeds
This is an excellent introduction to seed growing by Rob Johnson, the founder of Johnny's Selected Seeds. Includes detailed information for each species on growing, pollination, harvesting, cleaning and preservation. Order from: Johnny's Selected Seeds, Albion, ME 04910

SEED & PLANT SOURCES
SPECIALTY CROPS

BAMBOO

A Bamboo Shoot
1462 Darby Rd., Sebastopol, CA 95472
(send large S.A.S.E. for catalog)

Bamboo Sourcery
666 Wagnon Lane, Sebastopol, CA 95472

Endangered Species
P.O. Box 1830, Tustin, CA 92681

Panda Products
P.O. Box 104, Fulton, CA 95439

Steve Ray's Bamboo Gardens
909 79th Place South, Birmingham, AL 35206

BONSAI

The best place to locate sources is through your local bonsai club. The two groups listed below will help you find local members.

Bonsai Clubs International
2636 W. Mission Rd. No 277, Tallahassee, FL 32304

American Bonsai Society
P.O. Box 358, Keene, NH 03431

For a wide variety of starter plants and seeds, write these sources:

Girard Nurseries
P.O. Box 428, Geneva, OH 44041

Maplewood Seed Company
6219 S.W. Dawn St., Lake Oswego, OR 97034
Sells over 100 varieties of exotic and ordinary maples, send S.A.S.E. for a current pricelist.

Thompson & Morgan
P.O. Box 1308, Jackson, NJ 08527
Offers evergreen and deciduous seeds.

DRIED FLOWERS
Johnney's Selected Seeds
Albion, ME 04910

Park Seed Co.
P.O. Box 46, Greenwood, SC 29648

SPROUTING EQUIPMENT
Sunsprout Systems
9137 Spring Branch Dr. No. 306, Houston, TX 77080
Sunsprout offers automated sprouting machines, washing and de-hulling machines, label despensers, bags and seeds. They also have a licensing program for growers to market sprouts under the "Sunsprout" name, which includes full marketing support and supplies.

International Specialty Supply
820 East 20th St., Cookeville, TN 38501
I.S.S. carries sprouting seeds, several brands of automated sprouting and rinsing machines, boxes and bags, labels and packaging machines. They also publish an excellent free newsletter "Sprouting World", that covers growing ideas, trends, new products, etc. Their current catalog is $2.

MUSHROOMS

Far West Fungi
P.O. Box 428, S. San Francisco, CA 94083
(free catalog)

Field & Forest Products
N3296 Kozuzek Rd., Peshtigo, WI 54157
(free catalog)

Fungi Perfecti
P.O. Box 7634, Olympia, WA 98507

Mushroompeople
Box 159, Inverness, CA 94937
(free catalog)

Northwest Mycological Consultants
702 NW 4th St., Corvallis, OR 97330
(catalog $2)

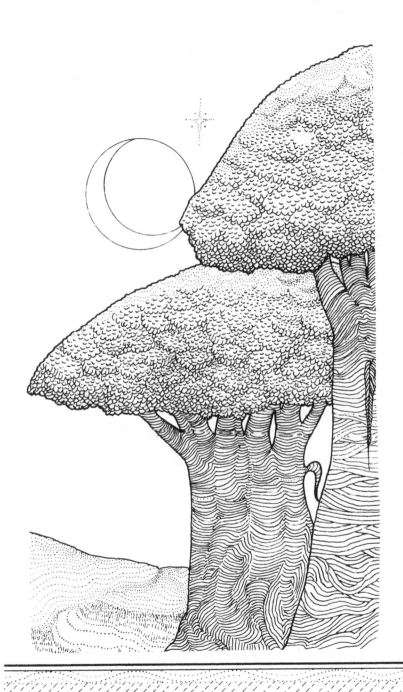

TREE CROPS

LANDSCAPING TREES

Even in a small suburban back yard, you can grow landscaping trees for profit. A large market exists to supply do-it-yourselfers and landscapers who need affordable medium-sized landscaping trees that can easily be transplanted.

For example, you can plant more than 300 seedlings in a fifty foot by fifty foot area. They will reach marketable size in three to five years. At today's prices you should expect an $8 to $15 profit per tree sold. Some growers specialize in "specimen" trees that are usually sold to landscapers to provide a focal point for a landscape plan. Profits of up to $100 per tree are not uncommon in this area.

Trees are an ideal backyard cashcrop for those whose spare time is limited. Unlike vegetables, trees don't require your constant attention. With occassional attention to proper pruning, spraying and feeding, trees almost take care of themselves.

Choose species that are popular in your area. Go to several retail nurseries to see what's in demand before you order seedlings or seeds. Unless you plan to specialize in a favorite species, plant a mix of trees. A popular assortment might include maple, oak, birch, ash, tulip, hemlock, pine, yew and arbor vitae.

Try to locate a wholesale seedling nursery in your area. If you can't purchase your seedlings locally, write for a catalog from the mail-order nurseries listed at the end of the chapter.

Since the selling season is short (early spring and late fall), you can concentrate your sales on a day that will produce the largest volume in the shortest time. One part-time grower runs a small display ad in the weekly paper for just four weeks in the spring and two weeks in the fall. He offers "Saturday Specials", and is only open from 10 to 4 those six Saturdays!

Another successful part-time grower avoids retail entirely, preferring to sell directly to landscapers who buy in quantity. A one page flyer, with a list of trees available and prices, is mailed monthly to landscapers, landscape designers and garden centers. He usually sells out his stock by mid-May each year.

"The best time to plant a tree was twenty years ago. The second best time is now."

CHRISTMAS TREES

An enterprising farm boy purchased 25 ducklings for 25 cents each. He let them run loose on the farm eating spilled grain and insects. He sold them in the fall for $1.25, a profit of a dollar each. He figured that, at this rate, if he bought a million ducklings the next spring, he would be a millionare by fall. Fortunately, his banker wouldn't lend him the $250,000 to buy the million ducklings!

The economics of growing and selling Christmas trees is similar to the duckling example, which is why Christmas tree farming is a popular and profitable specialty crop for more than 15,000 growers in the U.S. Over sixty million trees are sold annually, for prices of $9 to $40 each at wholesale and $16 to $80 at retail.

If you own or can lease as little as an acre of land, you can grow Christmas trees for profit. The trees normally need six to ten years to reach marketable size, depending on the species. The three most popular species of plantation-grown trees are: Scotch Pine, Douglas Fir and Balsam Fir. Scotch Pine, the leader, is preferred by more than half the retail buyers.

Scotch Pine is grown mostly in the midwest and northeast, while the northwest supplies most of the Douglas Fir, the second most popular.

Recommended plantings range from 1000 to 2000 seedlings per acre. With a special planting tool available through most seedling nurseries, an average person can plant 200 to 400 seedlings in just a day. Before planting, talk with your county extension agent about soil tests and advice on which varieties will do best on your land. Also check out market requirements. Visit several Christmas tree lots in season and determine prices and varieties.

"Living" Christmas trees are gaining in popularity, because the tree can be used again or transplanted to the yard after the holidays. They are more expensive, because digging and wrapping the root ball takes more labor.

When your trees have reached marketable size, you will need to decide how best to sell them. If you plan on wholesaling, you can sell through brokers who will either cut, bundle and ship your trees, or specify delivery.

Another option for you is to contact the retailers at their lots the year before you go to market. If you have a truck and the time, you can sell your trees for a higher price delivered to the retail lot.

If your tree farm has good road access, consider a "choose-and-cut" operation. Many counties and newspapers now publish free maps in the late fall, showing all the Christmas tree farms in the area. Your county extension agent can give you more details.

Ads in the local paper and a sign at the entrance to your tree farm will also generate a steady stream of customers. Use a simple "per-foot" pricing system. Have several inexpensive bowsaws ready for those who want to cut their own, and lots of twine to tie the trees.

Perhaps the most imaginative grower/seller of Christmas trees is Cathy Berwind, in Washington state. She works out of her basement, sending containers of trees and holiday boughs to Hong Kong! Each seven or eight foot tree sells for $150, and a simple wreath brings $35. Buyers include transplanted Americans, clubs and hotels.

Before you plant a single tree, talk to the growers associations, both local and national. Write to the National Christmas Tree Association, 611 East Wells St., Milwaukee, WI 53202 or telephone them at (414) 276-6410 to join or get the name and address of the secretary of your state association. The state associations usually stress tree production, while the national association emphasizes tree marketing. The national association also publishes "American Christmas Tree Journal."

HOLLY

The familiar red berries and glossy green leaves of the holly are usually associated with the traditional Christmas celebration. Yet the use of holly dates back as far as the Druids of ancient England, who believed the tree was sacred and used sprigs to decorate their homes. The holly tree has long been grown as an ornamental tree, hedge or groundcover.

For years, the annual prunings, sold to florists or retailed, have provided a lucrative holiday cash crop to many growers. In most areas, holly cutting and packaging begins in November as the berries mature and reach a full red color. The trimmings, called "sprays", are cut twelve to twenty-four inches long with hand shears and pole pruners. The sprays are sorted by size, with the smaller ones packed into "retail" cartons and the larger sprays into "bulk-florist" cartons.

As demand for holly has increased over the years, growers have planted commercial orchards, up to 100 acres, with holly trees. One five acre orchard in northern Florida, planted in American holly, averages 3000 pounds of cuttings each year. At a wholesale price of over $2 per pound, you can see the potential in holly.

Once you've planted hollies, it will take at least five years before they will produce a full crop of sprays. During that time, the young trees will require little attention. When the trees are at least five years old, the trimming can begin, but is usually limited to five pounds per tree at first. When harvesting sprays, always trim the tree so the natural shape is maintained, and remove damaged or low branches that block air circulation around the tree.

Before ordering your planting stock, check with prospective buyers, such as florists, to find out preferred holly varieties. Concentrate on local marketing as your orchard matures, to avoid the need for expensive preservatives or refrigeration. Contact local florists, for example, in the summer to pre-sell your holiday crop.

TREE CROPS
RECOMMENDED READING

The Tree Farm, by Robert Treuer
Little, Brown & Co., 34 Beacon St., Boston, MA 02106

**Christmas Trees for
Pleasure and Profit**
*Written to benefit both the novice
and the experienced grower, this
book guides the reader through the
stages of establishing and maintaining a stand of Christmas trees.
Available from:*

Musser Forests, Box 340-M,
Indiana, PA 15701

TREE CROPS
SEED & SEEDING SOURCES

American Holly Products
Route 49, Box 754, Millville, NJ 08332

Maplewood Seed Co.
6219 S.W. Dawn St., Lake Oswego, OR 97034

Musser Forests
P.O. Box 340-M, Indiana, PA 15701

National Arbor Day Foundation
Arbor Lodge 100, Nebraska City, NE 68410

Recor Tree Seeds
9164 Huron St., Denver, CO 80221

Carter Seeds
475 Mar Vista Dr., Vista, CA 92083

Carino Nurseries
P.O. Box 538, Indiana, PA 15701

Fowler Nurseries
525 Fowler Rd., Newcastle, CA 95658

Loucks Nursery
P.O. Box 102, Cloverdale, OR 97112
(Japanese maple specialist. Catalog $1)

Seed Source
Rt. 2, Box 265B, Asheville, NC 28805
(catalog $1)

Western Maine Nursery
36 Elm Street, Fryeburg, ME 04037

TREE FRUITS

APPLES

The apple is sometimes called the "King" of fruits, and rightly so. It is the most important commercial fruit crop of the temperate zone, grown world-wide from Siberia and Alaska to Georgia.

In our grandparent's time, a rainbow of varieties was grown, like the **Rhode Island Greening,** a cooking apple developed in 1748, **Blanc D'Hiver,** the gourmet apple of France and **Ashmead's Kernel,** a golden brown winter russett dating back to the year 1700.

Today, there is renewed interest in the old-fashioned apples, and new orchards are springing up across the country that specialize in the "antique" and gourmet apples. In northern California, Terry and Carolyn Harrison have established a small 1-1/2 acre orchard and nursery featuring the old-fashioned varieties.

Even though their orchard is at the end of a mile-long gravel driveway, they sell out their stock every year, and even sell many trees in advance of the spring planting season. To boost off-season sales, they have four one-day apple tastings in October. Visitors can sample slices of dozens of old-time varieties. If a visitor likes a particular variety, they can place an order and pay for the tree for spring delivery.

The Harrisons also belong to the Sonoma County Farm Trails producer's group. The 150 member group distributes a farms map and promotional materials to encourage people to visit the farms and buy direct.

HI-VALUE APPLES

Jill and Tom Vorbeck, Illinois apple growers, are selling their apples for up to $1.50 each! Of course, these aren't just "ordinary" apples. They are antique apples, like **Esopus Spitzenberg,** Thomas Jefferson's favorite apple. Or **Calville Blanc d'Hiver,** the favorite dessert apple of France that was grown for King Louis XIII in the 17th century. Some of the over

100 other heirloom apples offered by the Vorbecks include: **Baldwin, Blue Pearmain, Smokehouse** and **Winter Banana.**

While the Vorbecks grow many of their own apples, their mail-order apple sampler business is growing so fast that they have to buy apples from cooperating growers around the country. Send for their "Applesource" catalog. Route 1, Chapin, IL 62628.

Sal and Betty Guidice, owners of Pine Brook Orchard in Colchester, Conneticut, designed their orchard for handicapped pickers. Apple picking is an autumn treat for many, but it is not easy for little people — children and the handicapped. Their dwarf trees are all six feet or less, with the emphasis on "the good ones being within reach", says Guidice.

Pam and Gary Mount operate a small family orchard in New Jersey and report that, because demand far exceeds supply, they must bring in apples from other orchards in the area. They grow thirty different varieties, with customers often calling in advance to reserve a bushel or two of Macouns, Petrals or Wealthies. They say the big advantage to having a great variety of apples is spreading out the harvest.

The Mounts have just planted an additional acreage in trellised dwarf trees. They chose popular apples like **Stayman, Winesap, McIntosh, Jonathan, Macoun, Prima,** and **Empire.** The varieties were chosen not only for taste appeal, but also because they won't all ripen at once, spreading the harvest out over a ten week season.

The Mounts say a beginning fruit grower would be crazy to plant standard size trees today. "Dwarf trees should produce two to three times per acre what standard size trees produce, and they begin producing in three to four years," Gary says. In addition to apples, the Mounts produce thousands of gallons of fresh cider each season, using a blend of apple varieties. Gary claims the Stayman Winesap is the only apple that makes a decent cider on it's own.

According to a national grower's group, NAFEX, the best tasting antique or gourmet apples are: **Jonagold, Mutsu, Arkansas Black, Holiday, Spigold, Granny Smith, Cox's Orange Pippin, Newton Pippin, Wagener, Empire, White Winter Permain, Melrose, Gala, Ashmead's Kernel** and **Northern Spy.**

DWARF FRUITING WALL SYSTEM

Horticultural researchers have developed a trellis system for apple orchards that results in much higher fruit production than conventional orchards. A permanent trellis at least ten feet high and containing three or four wires (12 gauge galvanized) is installed at planting time.

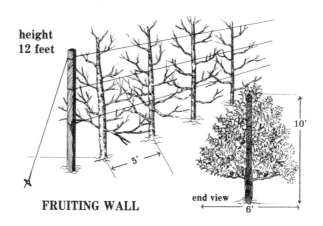

height 12 feet

5'

10'

FRUITING WALL

end view

6'

The trellis requires strong treated posts set three feet into the ground and anchored to a "dead-man" at each end. The posts are set 40 to 50 feet apart in the row, with rows about 14 feet apart. The dwarf apple trees are then planted six feet apart in the row. A permanent drip irrigation system is installed at planting time.

Once the trees begin bearing, usually in the third or forth year, the excess side limbs are thinned to allow sunlight to penetrate. The trees are pruned to a "Christmas Tree" taper. As the tree grows, the trunk is attached to the trellis wires. Mature orchards using this system report average yields of 50,000 to 60,000 pounds per acre!

In addition to high fruit production, growers are finding that picking and pruning are much easier and quicker, since no tree is over twelve feet tall.

FRESH CIDER

Many growers are turning their "wind-fall" and culled apples into freshly pressed cider to sell. Goodnature Products, (P.O. Box 233, East Aurora, NY 14052) manufactures modern grinders and presses for the small operator. Their "Roadsider" press and grinder system takes only sixty square feet of floor space, yet can produce up to 50 gallons of fresh cider per hour. At a current price of $2.50 per gallon for fresh pressed cider, you can see the potential. Many pick-your-own orchards also offer pressing or rent press time to their PYO customers.

APRICOTS

The apricot originated in China over 2000 years ago. It had reached Rome by the first century AD, and was brought to the U.S. by early settlers. Today, Russia is the largest producer of apricots. In the U.S., most commercial orchards are on the west coast, but new plantings are being established in the rest of the country.

Much of the reason for it's increased popularity is the superb taste of a tree-ripened apricot. Commercially grown apricots are picked while still green and firm. Since the apricot does not get sweeter or riper once it is picked, most commercial apricots taste terrible!

Once you've tasted a tree-ripened apricot, picked at the peak of flavor, you're usually hooked. So to be sure of premium prices for your apricots, plan to specialize in U-Pick or offer only tree-ripened fruit picked daily. Your customers will come back year after year for your "peak of flavor" apricots.

ASIAN PEARS

What is round and crunchy like an apple, crisp and juicy when ripe, yet with the flavor of a pear? It's an Asian Pear, the delicious fruit from the orient that's rapidly gaining in popularity from coast to coast. Back in Gold Rush days, Chinese miners brought the seeds to California and planted them along the streams of the Sierra Nevada. Today, the descendants of those first trees are producing commercial crops each year.

The Asian Pear has three apple-like qualities that make it appealing to would-be growers. It will keep up to two weeks at room temperature, and several months in cold storage. Second, unlike European pears, it ripens on the tree and is ready to eat when you pick it. Another advantage is that, like dwarf apples, the Asian Pear starts bearing early, usually by the third year.

Popular varieties include **Twentieth Century,** (also called Nijiseiki) the most popular in Japan, with a tender skin and mild juicy fruit; **Hosui,** a russeted and very productive variety that is one of the leading commercial varieties in California; **Shinko,** a sweet variety that keeps extremely well, and **Shinseiki,** with a fruit similar to Twentieth Century, but with a yellow skin and a hint of tartness.

Most commercial orchards are concentrated on the West Coast, where the Asian pear was first introduced, but the fruit can be planted in almost any area where apples and European pears can be successfully grown.

Asian pears are grown mainly for the fresh market and are selling for premium prices in the marketplace because of limited supply. In addition, there is export demand for the fruit, because countries like Taiwan cannot produce enough for their large population. In Asia, giving a pear to your host or boss has the same meaning as flowers or a bottle of wine in this country. Prices there run as high as $10 per pear!

In Japan, because of limited orchard space, Asian Pears are grown in a trellis system similar to that used for dwarf apple trees. In this country, with more available land, many growers are using a hedgerow system with the trees 7-8 feet apart in rows 14 feet apart. Average yield is 12,000 to 20,000 pounds per acre.

CHERRIES

Growers who are planning an orchard for "U-Pick" or direct sales should include cherry trees in their orchard plans. Because cherries ripen sooner than most orchard crops, your orchard income can start a month or two sooner.

Cherries are not suitable to all climates, and seem to do best in an intermediate climate. There are many types and species, but your main concern should be the suitability to your paticular climate. Check with your local county agricultural extension agent and local nurseries for specific recommendations.

The two types of cherries are "sweet" and "sour" (also called pie cherries). Although the sour cherries are easier to grow, usually there is not as much demand for them, so your orchard should have more sweet cherries than sour. For ease of picking and for ease of covering a tree in bird netting, use either genetic dwarf trees now available, or use an "interstem" dwarf.

Here's how an interstem dwarf works: A standard cherry rootstock is used because it tolerates a wide variety of soil and growing conditions. A sour cherry interstem, which is naturally dwarfing, is grafted on the rootstock. After a year, a sweet cherry variety is grafted to the interstem. The interstem method produces a hardy, adaptable and productive tree that tops out at 15 feet or less, making for easy picking and bird protection.

PEACHES

If the apple is the "King" of fruits, then the peach is certainly the "Queen". It is the second most important commercial crop in the U.S. and is grown from Florida to Minnesota. The first trees were planted during the 1500's by Spanish explorers near

St. Augustine, Florida, and spread as far north as Pennsylvania by Indians by the time the first English colonists arrived. By 1790, commercial orchards were thriving around Philadelphia.

Peaches are either clingstone or freestone. The clingstone variety, as it's name implies, clings to the seed and has to be cut away, but the freestone separates easily from the seed. The old-fashined clingstone varieties are more disease and pest-resistant than the freestones, but not quite as easy to process or eat.

Although peaches can be grown farther north, the southern half of the U.S. is generally considered prime peach country. Clingstone peaches are now enjoying a revival in the south, because health-concious buyers are willing to pay a premium price for unsprayed fruit. Several commercial sources for the old-fashioned varieties are listed later in the chapter.

PEARS

The pear is one of the most important commercial fruits of the world, grown extensively in Europe, Asia and the U.S. The most popular pears are the dessert pears that ripen in the fall.

In the U.S., over half the crop goes into cans. In Europe over half the crop is used to make "Perry", an alcoholic beverage! This traditional beverage is made with fermented pear juice and sometimes a bit of crabapple juice for flavor. After squeezing and straining, much like apple cider, yeast is added for fermentation. Then it is racked and bottled, to be sold and enjoyed later.

Commercial growers should focus on the dessert pears that bring premium prices, and winter pears for storage and canning. Here are some varieties to consider:

Anjou — large juicy fruit — best eaten fresh.
Bartlett — most widely grown commercially — sweet and tender fruit.
Comice — large green aromatic fruit, thought to be the best winter pear.
Seckel — superior dessert pear, sweet, creamy-white flesh.

PERSIMMONS

The Oriental or Japanese persimmon has been widely grown in China and Japan for many centuries. Starting in the late 1800's, the U.S. Department of Agriculture introduced several commercial varieties to this country. However, most of these varieties were astringent (makes your mouth pucker), when not fully ripe, and have since been displaced by newer non-astringent varieties.

In Japan today, two-thirds of commercial persimmon production is the two favorite non-astringent varieties, **Fuyu** and **Jiro.** Many orchardists in the U.S. are predicting that the Fuyu and Jiro persimmon will be the most popular "new" American fruit of the future.

In addition to being tasty and colorful, the persimmon trees live up to 100 years, and are adapted to climates too cold for citrus and too warm for good apple production. Current commercial yields average about 5 tons per acre.

Besides being eaten fresh, persimmons are widely used in cookies, cakes, pies, ice cream, jelly and jam. In the Orient, dried persimmons are most commonly sold whole, covered with sugar crystals that form as the persimmon dries.

PLUMS

Plums are available in a wide range of sizes, colors and tastes. The three groups most widely grown are the European plums, (blue and oval) Japanese plums, (red and round) and the newer hybrids that combine the flavor of Japanese plums and the hardiness of European plums.

The enormous number of varieties allow you to closely match proper varieties to your soil and climate. Some growers choose the varieties of European plum with an extra high sugar content to allow sun drying to make prunes. In the U.S., the plum/prune is second only to the peach in commercial stone fruit production.

Because there are so many varieties, check with your county extension agent or fruit specialist for regional recommendations.

TROPICAL FRUITS

When the average American thinks of fruit, it's usually apples, pears, cherries and peaches. But there's a whole new world of fruits that are almost unknown in this country, yet are becoming more available every day. The "edible landscaping" concept has also increased interest and awareness in many obscure fruit varieties. Most of these fruits are tropical or semi-tropical plants, and so require a mild climate for successful cultivation.

Gene Joyner, extension horticulturist in Palm Beach County, Florida, puts on a tropical fruit show every year, featuring over 150 species of fruit. Behind his office is the Mounts Horticulture Learning Center, a 3 acre botanical park that contains some 200 species of fruit. Says Joyner, "People need to see the plants before they plant. You can get a lot more fruit from a small area if you plant small trees, like carambola." Write him for a free copy of "Fruit Plants for South Florida", Palm Beach County Extension Service, 531 N. Military Trail, West Palm Beach, FL 33415-1395

Here are some of the more promising tropical tree fruits to grow:

Cherimoya — One of the world's most delicious fruits. The trees grow to just 20', start bearing in about five years, producing one pound green fruits with a creamy white flesh. Tastes just like a combination of pineapple, banana and strawberry!

Guava — This musky-flavored fruit is used for jellies and juices as well as eaten fresh. The small trees can be grown anywhere citrus thrives. The Pineapple Guava, also called Feijoa, is as new and exotic to the American market as kiwi was just twenty years ago. In New Zealand, the export demand alone has exceeded supply, and fruit experts feel the same will be true for U.S. growers.

Loquats — This moderate size tree produces large crops of 3" orange fruit. It is now commercially grown in California and Florida.

Lychee — Also called Litchi, these plants form a family of several varieties, including the *Longan*, *Rambutan*, and *Pulasan*. The Longan, smallest of the family, produces small 1" fruits, and is commercially grown in Florida. The Lychee, considered by many to be among the world's tastiest fruits, looks much like a large strawberry.

Mango — The mango is as popular in Asia as the apple is in the U.S. The fruit can be eaten alone, or used in pickles, preserves, jellies, curries and the famous mango chutney. Trees grown in the U.S. produce an abundance of pear-size fruit, but will tolerate no frost.

Papaya — Called the "tree melon" because it's fruit resembles a sweet melon. Available in tropical and subtropical varieties, it grows to a modest 20' tree.

Star Fruit — also called *Carambola*, this native of Indonesia is becoming more common in American markets, as it is now grown commercially in Florida.

Sapote — This tree produces a round yellow fruit that tastes much like vanilla ice cream. Orchards in California and Florida supply the growing demand for this fruit, long a Cuban favorite. The black sapote is called chocolate pudding fruit, because the flesh closely resembles a chocolate dessert.

TREE FRUITS
RECOMMENDED READING

Establishing and Managing Young Apple Orchards USDA Publication No. 1897

Available through your county extension agent.

International Dwarf Fruit Tree Assoc. c/o Department of Horticulture, Michigan State University, East Lansing, MI 48823

An association of growers and researchers which promotes research and distributes information about growing dwarf fruit trees. Publishes "Compact Fruit Tree" bulletin and sponsors orchard tours.

New York State Fruit Testing Cooperative Association P.O. Box 462, Geneva, NY 14456

This organization propagates and makes available to it's members new fruit cultivars from the agricultural experiment station. An annual catalog lists varieties for fruit trees, scionwood for grafting, and rootstocks available to members. Annual meeting features a large exhibit of new varieties, talks and conducted orchard tours.

North American Fruit Explorers c/o Mary Kurle, Treasurer. 10S 055 Madison Street, Hinsdale, IL 60521

An association of hobbyists and professional growers and plant breeders. Services to members include a quarterly magazine, the Pomona, and an excellent mail-order lending library. Members participate in fruit testing groups to evaluate cultivars of a wide variety of fruits.

State Cooperative Extension Service
Your local county extension office has a catalog of helpful publications for fruit growers. They can also refer you to the state extension fruit specialists for technical information

Modern Fruit Science, by Dr. Norman Childers. Horticulture Publications, Rutgers State University, New Brunswick, NJ 08903
This is a how-to book which has been an industry standard for many years, and is updated regularly. The book contains 960 pages covering planting, pruning, grafting, thinning, harvesting, storage and marketing.

Asian Pear Varieties — Pub. No. 4068 University of California — Ag. Sciences

Growing Persimmons — Pub. No. 21277 University of California — Ag. Sciences
Both the publications listed above and many others are available from: Agricultural Sciences Publications, University of California, 1422 Harbor Way South, Richmond, CA 94804

TROPICAL FRUIT ASSOCIATIONS

Rare Fruit Council International 13609 Old Cutter Rd., Miami, FL 33158
Offers monthly newsletter, plant exchanges, and an annual membership list.

California Rare Fruit Growers Fullerton Arboretum, C.S.U.F., Fullerton, CA 92634
Publishes a quarterly 36 page newsletter, annual yearbook, seed service, and experts to answer questions.

Indoor Citrus and Rare Fruit Society 176 Coronado Ave., Los Altos, CA 94022
Membership includes a quarterly newsletter, plantfinder's service, seed service, and discounts on books.

TREE FRUITS SEEDLING SOURCES

Bear Creek Nursery
P.O. Box 411-P, Northport, WA 99157

C & O Nursery
P.O. Box 116, Wenatchee, WA 98807

Fowler Nursery
525 Fowler Road, Newcastle, CA 956858

Interstate Nurseries
Hamburg, IA 51644

Lawson's Nursery
Route 1, Box 472, Yellow Creek Rd., Ball Ground, GA 30107

Henry Leuthardt Nurseries
P.O. Box 666, E. Moriches, NY 11940

Henry Morton Nurseries
Route 1, Box 203, Gatlinburg, TN 37738

New York State Fruit Testing Cooperative Association
P.O. Box 462, Geneva, NY 14456

Preservation Apple Tree Co.
Box 279, Mt. Gretna, PA 17604

Raintree Nursery
391 Butts Road, Morton, WA 98356

Southmeadows Fruit Gardens
Red Arrow Highway, Lakeside, MI 49116
Perhaps the largest selection of unique and antique apples, with over 200 varieties listed. In addition to a free price and variety list, they publish a 110 page book with an in-depth description of each variety.

Stark Brothers Nurseries
Louisiana, MO 63353

Orchard Equipment & Supply Co.
Route 116, Conway, MA 01341
Offers a complete line of cider-making supplies, as well as other orchard equipment.

TROPICAL FRUIT NURSERIES

Garden of Delights 2018 Mayo St., Hollywood, FL 33020
"Try before you buy" seems to be the motto of Murray Corman, owner of this unique nursery. He offers fruit from most of the plants sold, so you can taste new fruits. Plants are available as seedlings, or in pots, all at quite modest prices. Catalog $2.

South Seas Nursery P.O. Box 4974, Ventura, CA 93004
Specialists in tropical and subtropical fruiting plants. Send S.A.S.E. for pricelist.

VEGETABLES

Growing vegetables for market is an ideal backyard cash crop. Your investment can be quite small, the profits are great, and the risks are small. Demand is large and growing, particularly close to towns and metropolitan areas.

The large grocery chains buy most of their fresh vegetables from Florida, California, and Mexico, and must pass the high transportation costs on to their customers. Also, that produce may have been sprayed with chemicals of unknown toxicity. City folks like high quality fresh vegetables and will go out of their way to buy them. Also, by buying direct, they save money over the supermarket prices.

You will have a big advantage over the typical large mechanized growers, because you can use intensive high-yielding growing techniques such as inter-cropping (planting several crops in succession, up to four a year in some areas). Crop yields using these intensive techniques can be awesome, when compared to conventional row-cropping. For example, John Jeavons of Ecology Action, reports average per-acre yields increasing to between four and six times the U.S. average by using intensive growing techniques. In addition, intensive techniques such as raised planting beds, closer plant spacing, and organic fertilizers can actually reduce energy and water use substantially. The Ecology Action test programs have also demonstrated that a skilled grower can make as much as $20,000 on just 1/8 acre of land! For more information about this exciting new way to grow crops, get the books listed in the recommended reading section.

Using the intensive growing techniques perfected by Ecology Action, grower Michael Norton started his own half-acre "micro-farm" in nearby Berkeley, California.

Today, his high-yielding half-acre is grossing nearly $200,000 per year, selling thirty varieties of edible flowers, over thirty varieties of lettuce, salad greens and "designer" vegetables to Bay Area restaurants.

Only twenty vegetables were selected, out of the dozens available, because they have proven profitability. You should plan to experiment with different vegetables on a small scale to find out which ones have the best potential for you. Yield information per acre is based on both the U.S. commercial average for standard row growing techniques, and intensive farming. As you will notice, intensive vegetable techniques can increase yields (and profits) dramatically.

ASPARAGUS

Asparagus is a valuable crop, and one of the first to emerge in the spring. It can be grown in most of the U.S. Because it is a perennial, the crowns you plant should last for up to twenty years with reasonable care. Asparagus prefers a sandy well-drained soil with lots of organic matter worked in. The plants take time to mature, so your first full harvest will not be until the third year. Because the early crops bring the highest price, use the new row covers such as "reemay" to warm the soil for earlier growth. Average commercial yields are 2000 pounds per acre, with intensive yields of up to 6000 pounds per acre. Average price per pound ranges from $2.00/pound early in the season to 50¢/pound late season.

SNAP BEANS

Snap beans are a popular fresh vegetable, and grow well on most soils. The seed can be planted just after the last frost, with succession plantings every two weeks to insure a continuous harvest. The pole type snap beans produce a crop over a longer period than the bush types, but require more work to trellis. Average commercial yields are 6000 to 7000 pounds per acre, while intensive yields can be 18,000 to 20,000 pounds per acre. Average price is 66¢ per pound.

BEETS

The beet is a cool season root vegetable that is quite frost tolerant. Soil pH is very important to good growth, and should be adjusted to the ideal 6.5 if necessary. Succession plantings will provide a steady supply to harvest. The Cylindra beets pro-

duce a much greater yield than the regular varieties. Average commercial yields are 10 to 15 tons per acre. Intensive yields are 16 to 18 tons per acre, with prices averaging 15¢ per pound (with tops).

BRUSSEL SPROUTS

Brussel sprouts are best transplanted into the garden. Start from seed in a greenhouse or coldframe in early spring and transplant to growing area in early summer. Once in the ground, sprouts are easy to grow. For succession planting, put your brussel sprouts in the space formerly used by early peas or lettuce. Brussel sprouts tolerate even heavy fall frosts, in fact it even improves the flavor. Commercial yields average 16,000 pounds per acre, with intensive yields up to 30,000 pounds per acre. Prices average 75¢ per pound.

CABBAGE

The cabbage is an important vegetable crop because it is cold-hardy and adapts well to most soils. Most growers start the crop as transplants, but direct seeding also can be used. It is a heavy feeder, and requires plenty of fertilizer. To insure a steady supply throughout the growing season, plant early cabbage varieties, then mid-season, finally late varieties for winter storage. Average commercial yields are 24,000 pounds per acre, with intensive yields of 60,000 pounds per acre possible. Average price is 12 cents per pound.

CARROTS

Carrots prefer a fertile sandy soil, but will tolerate most growing conditions. To insure a good market appearance, be sure the soil is loose and free of rocks. Plant a succession of crops from early spring to mid-July in most areas to insure a steady supply. The demand for "baby" carrots is growing. So be sure to test one or two of those varieties, such as "Minicor". Average commercial yields are 40,000 pounds per acre, with intensive yields of 48,000 pounds per acre. Prices average 16-20 cents per pound.

CAULIFLOWER

Generally, cauliflower is grown like cabbage. It requires "sweet" soil and ample moisture. Most commercial growers start their early crop with transplants started in the greenhouse 4-6 weeks earlier. The transplants can be set out after the last frost. When the head is medium-sized, it should be blanched (protected from sunlight) by tying the outer leaves over the head with a rubber band. This will prevent head yellowing. Average commercial yield, with two crops per year, is 12,000 to 14,000 pounds per acre, with intensive growing yielding an average 32,000 pounds per acre. Prices average 50¢ per pound.

CELERY

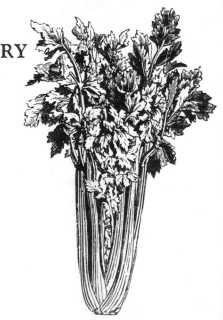

Celery is not hard to grow, contrary to popular opinion, if you have enriched soil and a constant supply of moisture. Start transplants 10-12 weeks before moving to growing area, and keep moist. Some growers clip and dry the excess leaves for seasoning (celery salt, for example). Average yield is 50,000 pounds per acre, with intensive yields of twice that possible. Prices average 18¢-20¢ per pound.

CORN

Almost everyone loves fresh-picked sweet corn. Although the profit per acre is lower than for other vegetables, it requires less work. So if you have enough land, and the equipment to till and cultivate it, by all means grow corn. Plant after the last frost and when the ground has warmed up, and every two weeks through early summer for a continuous crop. An alternative to succession planting is to use early, mid-summer, and late varieties at the same time. Average commercial yield is 12,000 pounds per acre. Prices average 70¢ per pound.

CUCUMBERS

Cucumbers are always popular as a fresh vegetable in salads and for pickling. They do well in most soils, and are normally seeded directly in the ground after the soil has warmed to 65 degrees. To get a head start on the growing season, many market gardeners start the seed in soil blocks three or four weeks ahead, then transplant. Be sure to provide ample water during fruiting to insure a large harvest. Average yield is 16,000 pounds per acre, with intensive yields of up to 150,000 pounds per acre! Prices average 25¢ per pound.

GARLIC

Garlic has been popular as a cash crop for years with market growers, sometimes even being called the "mortgage lifter". Modern medical research has confirmed the beneficial effects of garlic in controlling high blood pressure. Many pet food manufacturers add garlic to discourage fleas. Some growers specialize in "Elephant" garlic because the giant bulbs are quite mild and preferred by many customers.

Garlic is usually started in early spring by pulling apart bulbs and using individual cloves for sets, much like onions. Average commercial yield is 10,000 pounds per acre, with intensive yields ranging up to 40,000 pounds per acre. Prices range from 70¢ to $2 per pound.

LEEKS

The leek is a non-bulbing type of onion with a long thick stalk. They prefer a rich soil, and should be started indoors and transplanted to insure the largest stalks. The stalks should be blanched during the growing season, by hilling up the soil around each stalk every two weeks. Some varieties can be mulched and left in the ground over winter to provide late season crops, or kept in dry storage for up to three months. Average commercial yields are 20,000 to 30,000 pounds per acre.

LETTUCE

Lettuce is quick, easy to grow, and always in demand. Plant three or four varieties of leaf or bibb lettuce, and plant succession crops each week for a steady supply. Head lettuce can be  hard to grow, and is not as profitable as the leaf varieties. The cos, or romaine lettuce is also a good seller. Since lettuce is hardy, plant as early as the soil can be worked. Most growers transplant their spring crop to get an early start, then direct-seed the later crops.

Mel Bartholomew, in his book "Cash from Square-Foot Gardening", has a proven plan for raising lettuce that yields $7 per square foot! Average commercial yields are 24,000 pounds per acre, with intensive yields of up to 64,000 pounds. Average price is 24¢ per pound.

OKRA

Okra is a popular crop, especially in the south, where it is called "gumbo". It will grow in almost any soil, but must have hot weather to grow well. The pods should be harvested when young and still tender. Average commercial yields are 8,000 pounds per acre, with intensive yields of 16,000 to 20,000 pounds per acre. Price ranges from 30¢ to 40¢ per pound.

ONIONS

Because onions are easy to ship and store, there is usually an abundant supply of standard onions available. Market growers should specialize in unique onions that are in demand and harder to ship and store. For example, the famous Walla Walla Sweet is a premium priced, extra mild slicing onion always in demand. Another specialty is the bulbless bunching onions, called scallions, that are always in demand as a fresh vegetable. Another popular "gourmet" specialty is French shallots, used for flavoring soups, salads, casseroles and other dishes. Be sure to plant the true French shallots, which have a pink tint to the skin and flesh. Average commercial yields are 32,000 pounds per acre, with intensive yields of up to 60,000 pounds per acre. Average price is 30¢ to 60¢ per pound.

PEAS

The newer stringless edible pod peas are in great demand today. The old standard oriental snow peas are the most popular varieties. The edible pod varieties are grown just like regular peas, planted as soon as possible after the ground thaws. The sugar snap variety is well-suited to intensive cultivation using a vertical trellis system. Average yields for intensive cultivation is 8,000 pounds per acre at $1.00 a pound.

PEPPERS

Sweet peppers, or bell peppers, should be started indoors 6 to 8 weeks before transplanting. Like tomatoes, they need a long warm growing season to do their best. A sandy well-drained loam is best, with ample water as the plants mature. Harvest when the peppers are full-size and firm. Leave part of the crop unpicked to turn red, for those customers who prefer a red pepper — usually for canning. You might want to try a few semi-

hot pepper varieties to test local demand. Average commercial yield is 10,000 pounds per acre, with intensive yields of up to 26,000 pounds per acre. Prices are about 40¢ per pound.

RADISHES

Radishes are quick and very profitable. They like a light sandy soil, and can be direct-seeded as soon as the ground can be worked. To insure a steady supply, plant another batch each week through mid-spring. Make sure your radishes get ample moisture. This will keep your radishes mild and tender. Try some specialty radishes for variety, such as "Easter Egg", with red, pink, purple and white bulbs. Kids love them! Another popular specialty radish, "Daikon", is a long white radish popular in oriental

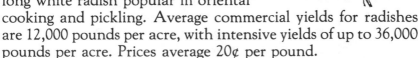

cooking and pickling. Average commercial yields for radishes are 12,000 pounds per acre, with intensive yields of up to 36,000 pounds per acre. Prices average 20¢ per pound.

SPINACH

Spinach prefers a moist fertile soil and a neutral pH to do well. It germinates well in cool weather, and can be direct-seeded as

soon as the ground can be worked. Many market growers plant spinach in late September, for an extra-early spring crop. Since it bolts in hot weather, schedule your plantings every ten days through late spring, then again in late July to mid-August for a fall crop. Average commercial yields are 8000 pounds per acre. Intensive yields are 16,000 to 24,000 pounds per acre. Average price is 30¢ to 40¢ per pound.

SQUASH

Summer squash is a fast-growing and profitable vegetable. The most popular varieties are the yellow straight neck and yellow crookneck. Be sure to pick bushy varieties, which are easier to maintain than the regular varieties. Most will yield a generous crop all summer if kept picked. Start indoors in soil blocks or peat pots three or four weeks before transplanting to growing area. Don't disturb the sensitive roots! A profitable specialty for some growers

is "novelty" squash, such as the **Red Kuri** Japanese squash, or the **Green Hokkaido,** or many of the unique squashes grown for ornamental gourds. Average yields are 18,000 pounds per acre, with intensive yields of up to 30,000 pounds per acre. Average prices are 15¢ to 50¢ per pound.

SWISS CHARD

Swiss chard prefers a sandy well-drained soil. Plant early-late spring after the last frost. As the plants mature, snip individual leaves near the base of the plant. New leaves will grow back and provide a steady harvest. Swiss chard is popular as a cooked green and for salads. Average commercial yield is 24,000 pounds per acre, with average prices of 25¢ per pound.

TOMATOES

Because tomatoes are one of the easiest crops to grow and require little land, every market gardener should grow them. The challenge is to have a tomato crop ready to go two or three weeks ahead of others, as the demand and the prices are highest early in the season. To get a head start, use transplants. You can double your tomato production by using tomato cages. In addition, growing and harvesting labor is greatly reduced. You can easily make your own using stock 6" concrete reinforcing mesh from the lumberyard. The cost is much less than the ready-made cages. Plant a diversity of tomatoes, including early varieties, paste tomatoes, and cherry tomatoes. Average yields are 24,000 to 50,000 pounds per acre, with prices averaging 30¢ to 40¢ per pound.

ETHNIC VEGETABLES

Twenty years ago, the average American supermarket stocked fifty items in the produce department. Today, that number has more than tripled, and new fruits and vegetables are being added monthly. Here's why . . . immigrants from all parts of the world have brought their native foods with them to North America. As others experience these new foods, demand increases and alert growers add or increase plantings of the newly popular varieties. While the plants listed here are all enjoying demand, in some cases it is limited to areas of large ethnic populations. So investigate your local markets before planting large quantities of any vegetable. Here are some of the more popular vegetables:

Asian Celery — Similar to the celery now sold at supermarkets, but grown to produce leafy greens with a strong flavor and aroma.

Black Radish — This import from Eastern Europe is as big as a turnip with white pulp and a black skin. They are used as a garnish or in salads.

Borage — in Europe, the leaves of borage are used as a salad green. The edible flowers are used as a garnish or in salads.

Burdock — Popular in Japanese cooking, where it is called "Gobo" and used in stir-frys. The Japanese claim that eating the root increases endurance and sexual virility.

Celeriac — In Northern Europe, these roots are used raw in salads, or cooked like carrots or parsnips in soups and stews.

Chicory — The chicory family includes endive, and has been popular in Europe for centuries. Now enjoying a surge in popularity in the U.S. as a "designer" vegetable. The fresh greens are used in salads and the roots are prepared like a parsnip. The dried and ground root is a popular coffee substitute.

Cilantro — Also called **Coriander,** this aromatic plant is used in many dishes, while the ground seed is a popular spice.

Daikon — This large white radish is widely used in Oriental dishes. Because daikon is much milder than common radishes, it is more widely used for stir-frys, stews or pickled.

Eggplant — As our Asian population increases, so does the diversity of eggplant colors and sizes. Eggplant has long been a major crop in Asia, where the colors range from white to green and yellow to the familiar purple. The striped Thai eggplants are popular, as are the long Japanese eggplants.

Finnocchio — Also called **Florence Fennel,** this vegetable is used both cooked and raw in Italian cookery much like celery.

Jicama — This crunchy sweet root is native to Latin America. It can be eaten raw or cooked, and is used in salads, soups and stir-frys.

Lamb's Lettuce — Also called **Corn Salad,** this vegetable is a well-known winter salad green in Europe.

Salsify — Also called **Vegetable Oyster,** this native of Europe produces a one foot long taproot, which is cooked like other root crops such as carrots. The roots are sometimes dried and ground like chicory to make a coffee substitute.

Yams — True yams are an important food crop in much of the world, and are used as a primary starch plant in Africa, China, India and the Caribbean. The yams found in most super-markets are not really yams, but a variety of sweet potato! Most true yams are potato sized, but some varieties routinely grow to 600 pounds!

MARKETING YOUR VEGETABLES

PICK YOUR OWN — Vegetables in supermarkets get more expensive every year with the growing cost of labor to harvest and pack the produce. In a P.Y.O. setup the harvesting and packing labor is supplied by your customers and they pay you for the privilege! Most P.Y.O. growers price their produce at 10 to 20 percent below retail. People will gladly pay that for fresh-picked local produce. For more complete P.Y.O. information, check out the chapter on marketing your crops.

RETAILING YOUR VEGETABLES — You can sell your produce directly from the garden, or if your garden is large enough, with a roadside stand. Many growers prefer to take their produce to market in order to reduce their selling time. Most areas have a Farmer's Market on Saturdays where you can rent a space to sell your crops.

ORGANIC VEGETABLES — There is an excellent market for produce grown without chemical fertilizers, herbicides and pesticides. A rapidly increasing number of people are aware of the potential problems caused by chemicals, and choose to avoid the risks. Growers who specialize in organic produce can charge a premium for their crops because of the extra labor in-volved.

The market for organic food has exploded in recent years, with estimated sales of organic food over $10 billion dollars last year. According to Betty Kananan, director of the Ohio Ecological Food and Farm Association, "The demand for organic food far exceeds the supply. I get calls and letters every day from the Cleveland and Columbus areas for organic vegetables and fruits. We need more organic growers, particularly vegetable growers."

In Texas, the Texas Department of Agriculture has become the largest promoter of organic produce in the state. The department is helping to establish the Texas Organic Growers Association to do co-op marketing of Texas-grown organic products.

"There's a $3 billion dollar a year market out there for interested Texas growers," says Jim Hightower, the Texas Commissioner of Agriculture. His department is launching a major promotional campaign to help sell the "TDA Certified Organic" label.

Check with your state's department of agriculture. Chances are good they have programs that can help you if you're growing organic.

SELLING TO RESTAURANTS – According to Mel Bartholomew, your local restaurants are a potential goldmine for backyard market gardens. According to Mel, "They buy everything you can deliver, and pay you top retail prices, in cash on the spot." He has spent much time and effort learning the needs of the restaurant trade and how best to sell your produce to restaurants. Read his new book "Cash from Square Foot Gardening", to learn all about this new approach to selling your vegetables for top dollar.

MIDGET VEGETABLES – Growing the unique midget vegetables in your garden can give you a cash crop with almost no competition. Growing midget vegetables is no different than growing the standard varieties, except that most midgets are quicker to mature.

Examples include midget carrots and beets, popular with women who do fancy canning. **Tiny Sweet** carrots, when mature, are only three inches long with a perfect cone shape. **Spinel** beets have a perfectly round golf ball size at maturity, just right for pickling. **Golden Midget** corn produces perfect four-inch ears in just two months. Markets for midget vegetables include individuals, restaurants, gourmet food stores, caterers and weekend farmer's markets.

BEST SPECIALTY VEGETABLES

For over 100 years, researchers at Cornell University have been growing and reporting on both old and new vegetable varieties, both for home gardeners and market growers.

According to Roger Kline, Research Director for specialty crops at Cornell, "The choice and diversity of vegetable varieties is exciting — we are like kids in a candy shop, not knowing what to choose first.

Perhaps the symbol of specialty vegetables should be the round, re-leafed, white-veined head of radicchio. It possesses the natural beauty and premium price of the most special of the specialty crops we've grown in our own test plots."

According to Kline, other "designer" vegetables that are popular now include: Chicory, edible-pod peas (both snow peas and sugar snap peas), fancy lettuces in green-red combinations, such as oakleaf. The tiny lettuces, such as "Tom Thumb", which make no more than two salads are back in favor again, as well as the colorful "Red Sails" and "Deep Red" loose-leaf varieities.

Colorful peppers that mature to red or yellow or chocolate are also in vogue, as are round zucchini, mustard greens and Romanesco broccoli.

To try the new varieties, you can order collections from Shepherds, such as "French Garden", "Italian Garden", and "Edible Flowers." (Shepherds Garden Seeds, 7389 West Zayante Rd., Felton, CA 95018 — catalog $1). Cornell has put together an extensive list of recommended specialty vegetable crops. For the current listing, send $1 and a large S.A.S.E. to: Specialty Crops Program, Department of Vegetable Crops, Cornell University, Ithaca, NY 14853.

VEGETABLES
RECOMMENDED READING

Pick-Your-Own Farming: cash crops for small acreages, by Wampler & Motes.

This 176 page book offers tips on growing the "16 proven money-making crops in the PYO operation." Both experienced growers and beginners should find this book very helpful and profitable.

Vegetables — How to select, grow and enjoy, by Derek Fell, HP Books.

This excellent full-color paperback is a complete step-by-step guide to growing more than 80 popular vegetables. A good basic book for the beginning market gardener.

Every year, the Horticulture Department at the University of Illinois holds a vegetable growers school, with dozens of workshops given by experts on vegetable production and marketing. One of the reasons it's so popular is that many of the workshops are given by the growers themselves, rather than school staff. If you can't attend, the proceedings of each yearly school are available for a modest $7. Write to: University of Illinois, Dept. of Horticulture, 1005b Plant Sciences Lab, 1201 South Dorner Dr., Urbana, IL 61801.

How to Grow More Vegetables — than you ever thought possible on less land than you can imagine, by John Jeavons

The Backyard Homestead Mini-Farm, by John Jeavons.

For over fifteen years, John Jeavons and Ecology Action have been developing their "Biodynamic/French Intensive" method of gardening. The latest refinement of the method has been to develop a "mini-farm" of 1/8 acre which, with experience, can provide an income of up to $20,000 per year working only eight months. The San Francisco Chronicle said it best: "The farm of the future . . . it's amazing . . . $20,000 a year and all you can eat!" Available from: Bountiful Gardens, 5798 Ridgewood Rd., Willits, CA 95490.

Knott's Handbook for Vegetable Growers, by Lorenz & Maynard.

Since the first edition over 30 years ago, this 400 page handbook has become a classic reference for all phases of vegetable production, from planting to marketing. The current edition has been revised and expanded to include the latest recommendations.

Cash From Square Foot Gardening, by Mel Bartholomew.

Another experienced intensive grower, Mel has developed a system to allow part-time gardeners to earn up to $5,000 each summer in their back yard.

VEGETABLE SEED SOURCES

Burpee Seeds
22345 Burpee Building, Warminister, PA 18974
Offers commercial seed catalog for market growers.

Farmer Seed & Nursery
818 N.W. 4th St., Fairbault, MN 55021
Carries many hard-to-find midget varieties.

Harris-Moran Seed Co.
3670 Buffalo Road, Rochester, NY 14624
Offers free "Commercial Vegetable Grower's Seed Guide."

Johnny's Selected Seeds
310 Foss Hill Rd., Albion, ME 04910
Specializing in vegetable seeds for northern growers — offers generous bulk-buyer discounts and personal service for market growers.

Nichols Garden Nursery
1190 N. Pacific Hwy., Albany, OR 97321
Extensive listing of vegetable seeds, including bulk starts for elephant garlic.

Redwood City Seed Co., P.O. Box 361, Redwood City, CA 94064

Specializing in unusual vegetables, such as 30 chili pepper varieties, Native American Indian vegetables from Hungary and the Orient. Catalog $1.

Seedway, Inc., P.O. Box 250, Hall, NY 14463
R.H. Shumway, P.O. Box 1, Graniteville, SC 29829

They have been selling vegetable seeds since 1870, and now offer the largest selection of open-pollinated vegetables in the U.S.! Catalog is $1.

Sunrise Enterprises, P.O. Box 10058, Elmwood, CT 06110

Sells seeds by the packet or by the pound for over 150 varieties of Oriental vegetables.

Tomato Growers Supply Co., P.O. Box 2237, Fort Myers, FL 33901

You asked for it! Over 160 different varieties of tomato seed, plus tomato growing supplies. Free catalog.

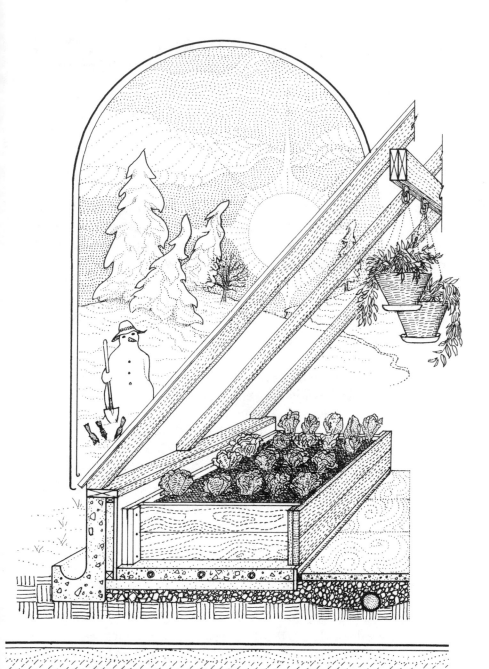

GREENHOUSE GROWING

Many backyard cashcrop growers build a greenhouse to start their own plants, then expand to growing vegetables, herbs and flowers for sale as bedding or potted plants. When you think about the possibilities of year-round growing, it's no wonder most serious growers use a greenhouse.

Most growers get started in greenhouse growing with bedding plants such as vegetable and flower starts, and then expand to flowering houseplants and potted foliage plants. According to the U.S. Department of Agriculture, the sales of bedding plants are growing at better than 15 percent a year. Best selling bedding plants are petunias, followed by impatiens, with tomatoes in third place. Other favorite bedding plants are geraniums, peppers, marigolds, begonias and pansies.

In addition to ornamentals, many greenhouse growers also grow vegetables ready to eat. The three most profitable greenhouse grown vegetables are lettuce, cucumbers and tomatoes, because they can be sold out-of-season to restaurants, quality grocers and individuals.

With heating costs rising every year, greenhouse growers are paying more attention to solar design when building a new greenhouse. Proper solar design can help reduce both heating costs and overheating on sunny days.

Because the plastic and fiberglass commonly used in greenhouse coverings are poor insulators, the areas that don't get direct sunlight, such as the north wall and roof, can be insulated to reduce heat loss. Then if a heat storage mass, such as rock, is put in the greenhouse, it can store excess day heat for use when the sun is not shining. This combination of reduced heat loss and increased heat storage can turn an ordinary greenhouse into a "season-stretcher" greenhouse and extend your growing season by several months in the spring and fall. To grow plants year-round, you will usually need to provide a supplemental heat source, depending on your climate.

SOLAR SITING

To allow the maximum amount of sunlight into your greenhouse, the long side should face as close to due south as possible. If your location does not allow a perfect exposure to the southern sun, remember that even if you face southeast or southwest, the solar energy will only be reduced by 25 percent.

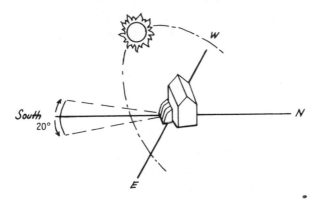

Percentage of solar gain available as the site moves away from true south.

True South	100%
20° East or West	95%
30° East or West	90%
45° East or West	72%

GREENHOUSE SIZE

When planning your new greenhouse, keep in mind your crops, climate and needs. Will yours be a year-round greenhouse or a season-extender? Growing crops to maturity in a greenhouse takes much more space than starting transplants in a flat. For example, Growers use a rule of thumb that each flat of 72 bedding plants or starts takes about 1 1/2 square feet of space in the greenhouse. Yet those same 72 plants, if they were tomatoes, for example, being grown for the off-season market, would take not 1 1/2 square feet but 360 square feet!

Experienced growers usually estimate how much space they think they will need, then double or triple it! To allow for expansion, can your greenhouse be stretched in length? To allow for easy expansion, many commercial greenhouses, called "gutter-connected", are designed to be added on each side, for cost and space efficiency.

GREENHOUSE COMPONENTS

Foundation — For smaller greenhouses, the preservative treated sill is often sufficient for reasonably level ground. Be sure to anchor the sill to avoid wind uplift problems. Standard screw-type mobile home anchors work well as hold-downs. Pressure-treated 4" x 4" wood posts set 18" to 24" in the ground on 4' centers make a simple and economical foundation also.

Another consideration involves taxes. Some tax assessors feel a permanent foundation of concrete means a greenhouse is a permanent structure and assessed much higher than a temporary greenhouse that can be dismantled and readily moved. A chat with your local assessor before you begin building could save you money on taxes.

For a permanent greenhouse, a poured concrete or block foundation is best. Be sure to add foamboard insulation to the outside before backfilling to prevent excess heat loss through the perimeter of your greenhouse.

Frame — Choose frame materials for long life, strength and cost. Most commercial frames use galvanized steel tubing for it's high strength and a wide range of shapes possible. Owner-builders who are building their first smaller greenhouse should consider wood, as it's easier to work with, readily available and inexpensive. Be sure to use wood that's been treated with a preservation suitable for contact with plants! A wood frame should be painted with an exterior primer and white paint to better reflect light in the greenhouse.

Glazing — Glass is the best material for greenhouse glazing, but quite expensive. In addition, it should be tempered for safety, and the extra weight requires heavier framing support.

Fiberglass, also called FRP, or fiberglass reinforced polyester, is a durable and modestly priced greenhouse glazing. It is also safer than glass, but transmits slightly less light. Only use greenhouse grade fiberglass, with a 10 year guarantee. The long-lived grades use a special tedlar coating to prevent yellowing from ultra-violet rays.

Polyethylene — Poly film is the covering of choice for most large commercial greenhouses, despite it's short life. The advantages are light weight and low cost. Most manufacturers of greenhouse poly use an ultraviolet inhibitor to increase the life of the film to up to three years. For year-round greenhouses in colder climates, two layers are generally used. This can reduce heat loss up to 40 percent. Most growers also pressurize the airspace between the two layers with a small fan to create a dead air space, which reduces heat loss even further.

HEAT STORAGE

Greenhouse researchers have found that a greenhouse must have enough "thermal mass" to store the excess day-heat for night-time use. They have also learned that the root zone is the crucial area for maintaining growth, so if the root zone soil can be kept warm, the plants will grow, even at 32 degrees! To keep the root zone warm at night or when the days are cloudy, the thermal mass beneath it can be warmed with excess heat during the day. To do this, the heat from the top of the greenhouse is pushed down with a fan and duct system and stored below the growing area, in crushed rock or sand, with standard 4" plastic perforated pipe in it to circulate the warmed air.

Beneath the rock, a layer of extruded polystyrene (also called blueboard) is used to keep the heat in the rock storage bed from being lost to the ground. Another approach is to build the growing benches on top of black 55 gallon drums filled with water. The sun warms the water during the day, and the heat is released at night.

VENTILATION

Greenhouses need plenty of ventilation for cooling, humidity control, and to restore carbon dioxide used by plants. Solar vents, available at most greenhouse suppliers, are automatic, requiring no electricity, and open and close according to the temperatures inside the greenhouse.

In the summertime, shade cloth should be used in conjunction with solarvents or a temperature-controlled electric exhaust fan can be used. The fans are available through the greenhouse suppliers listed at the end of this chapter. In addition to shade cloth, most suppliers offer spray-on shading compounds that can be washed off in the fall.

SUPPLEMENTAL HEAT

If you plan to use your greenhouse through the entire winter, you will need supplemental heat in most climates. While there are many types of air-type heaters available using electricity, propane or kerosene, the newest greenhouse heaters use circulated hot water to heat from below. The natural convection heating of hot water heating is more cost-effective than heating the air in the greenhouse.

As mentioned earlier, most plants benefit more from having the root zone warmed. A simple system for smaller greenhouses, using a stock household water heater, and in-line pump, and plastic pipe under the growing area can provide a low cost supplemental heat system. For larger greenhouses, the same system, using a boiler and finned tubing for more efficient heat transfer, is saving commercial growers up to 50 percent of their previous heating costs, and is available from most greenhouse suppliers.

GREENHOUSE CROPS

Having a greenhouse will expand your growing possibilities immensely. You will be able to extend the growing season and multi-crop, grow vegetables off-season, produce bedding plants or annuals for spring sale, grow holiday plants such as pointsettias for Christmas; even grow tropical plants in the dead of winter.

Of course, the plants you choose to grow will be controlled by the conditions you can afford to maintain in your greenhouse. Local climate, such as extremely cold winters or extended overcast periods such as are found in the maritime regions will influence your plant choices. But paying attention to the internal climate zones in your greenhouse can make a big difference. For example, heat loving plants can be grown up near the peak of your greenhouse, the shade tolerant plants close to the ground and the light loving plants near the middle.

BEDDING PLANTS (ANNUALS)

Bedding plants are started in flats in your greenhouse. Most are started in early spring for summer blooming, but many can also be started in the fall for winter or indoor flowering.

Here are some of the more popular annuals, with the time to plant in the greenhouse for each plant's blooming season. Remember that there are many other annuals to experiment with — these are just the more popular varieties. Be sure to check the flower chapter for more ideas.

Ageratum — Plant in February-March for summer blooms or August for spring blooms.

Alyssum — Plant in March for summer blooms or August for winter blooms.

Aster — Plant in February for summer blooms, August for winter blooms, and December for spring blooms.

Calendula — Plant in January for spring blooms, March for summer blooms and August for winter blooms.

Candytuft — Plant in January-February for summer blooms and in the fall for spring blooms.

Cosmos — Plant in March for summer blooms and October for spring blooms.

Heliotrope — Plant in February for summer and July for winter blooms.

Lobelia — Plant in February for summer blooms, September for winter blooms and December for spring blooms.

Marigold — Plant in February for spring blooms, April for summer blooms and August-September for winter blooms.

Pansy — Plant in March for summer blooms, July for winter blooms and December for spring blooms.

Petunia — Plant in March for summer blooms and November for spring blooms.

Phlox — Plant in March for summer blooms and September for spring blooms.

Portulaca ("Mossrose") — Plant in March for summer blooms.

Snapdragons — Plant anytime for flowers in three to four months (longer in winter).

Zinnias — Plant in April for summer blooms and August for winter blooms.

BLOOMS

A greenhouse will allow you to produce early blooming flowers for sale. Some of the hardy bulbs like the crocus, hyacinth and tulip need a cold period to simulate winter. You can put the hardy bulbs in a refrigerator, or buy pre-conditioned bulbs which are ready for forcing into early bloom. The more popular are listed below, together with the planting/blooming times.

Amaryllis — Plant in September/October for February blooming.

Crocus — Plant in September/October for December/January blooms.

Cyclamen — Plant in August for winter blooms.

Daffodil — Plant in September through November for blooms December through March.

Freesia — Plant August through November for blooms December through April.

Hyacinth — Plant September through November for blooms December through February.

Iris — Plant November through January for blooms February through April.

Narcissus — Plant September through November for blooms December through March.

Tulip — Plant in September through October for January/February blooms.

HERBS

Use your greenhouse as a season extender for herbs. Start herbs early for late spring and early summer sale in 2" to 4" pots. Culinary herbs are generally the best sellers, as many customers prefer to cook with fresh herbs, but don't have the patience for seeds to grow into a usable plant! Popular culinary herbs for spring/ summer sale include **Basil — Chives — Dill — Oriental Garlic (Garlic Chives) — Marjoram — Mint — Oregano — Parsley — Rosemary — Sage — Savory — Tarragon — Thyme.**

HOUSEPLANTS

The market is expanding fast for both foliage plants and flowering plants. Experiment with the proven houseplants listed below to determine local demand. Tour local greenhouses and plant nurseries to see what's selling before making a major commitment of time and money to any plant. Refer to the books listed at the end of this chapter for specific growing information on any of these plants. **African Violets, Begonia, Cactus, Chrysanthamum, Coleus, Ferns, Fuchsia, Geranium, Impatiens, Ivy, Jasmine, Orchid, Poinsettia, Primula, Spider Plant.**

VEGETABLES

There has been a tremendous increase in the quantity of vegetables grown in greenhouses in the last few years. Shortages of good land near metropolitan areas, water scarcity, and the rising cost of transporting vegetables from growing areas to distant markets have all had a strong influence.

Many market vegetable growers are discovering the benefits of greenhouse production. Greenhouse vegetables can be grown intensively to produce a larger crop in less area. A good example of this is the vertical trellis system used by tomato growers to allow the tomatoes to climb up for increased light and growth.

Because evaporation and surface runoff can be controlled, the amount of water required to grow a greenhouse vegetable crop is a fraction of the water required for a field crop of the same vegetable. This is an important consideration as water shortages, water rationing and water contamination become more and more widespread.

In a controlled environment greenhouse, two or three and sometimes four crops can be grown over the entire year, instead of just the one crop usually possible with field-grown produce. This allows a grower to produce a steady income year-round rather than being dependent on banks for seasonal production loans and seed loans. Producing a crop out-of-season can also mean much higher prices for your vegetables, since they don't have to compete with local field-grown crops.

Because a greenhouse operation takes less land than a field-grown operation, you can locate your greenhouse closer to metropolitan markets, where the demand is greater for your crops. Your transportation costs and delivery time can be reduced also.

The weather has been more and more unpredictable in the last few years. Some long-range weather forcasters claim that it's going to get worse, and that the extremes of cold and drought and grasshopper invasions will be more frequent. Inside the greenhouse you can control the climate to insure that your crops are not adversely affected if the weather takes a turn for the worse.

And speaking of grasshoppers, most greenhouse growers agree that pest and disease management is far easier in the greenhouse than in the field. This means increased yields, reduced expenses for chemicals, and healthier produce for your customers.

The vegetables listed below are by no means the only ones possible as greenhouse crops, they are the established proven sellers. Experiment on your own, and by talking to other growers to find out what's best for your market and climate. Also get a copy of the Johnny's seed catalog. They list specific varieties of vegetables based on growing experience, which are especially suited to greenhouse production because they are productive in a compact space and at cooler temperatures.

COOLER GREENHOUSE
(35 to 60 degrees)

Beets – Chinese cabbage – Carrots – Endive – Chinese greens – Kale – Kohlrabi – Lettuce – Bunching onions – Shallots – Parsley – Radish – Spinach – Swiss chard.

WARMER GREENHOUSE
(60 to 90 degrees)

All vegetables listed above, plus Celery – Eggplant – Cantaloupe – Peppers – Tomatoes.

GREENHOUSE RECOMMENDED READING

The Complete Greenhouse Book, by Peter Clegg and Derry Watkins.

This innovative and practical book shows you how to plan and build a solar greenhouse; what to grow and how to grow it.

Solar Greenhouse for the Home, U.S.D.A. No. NRAES-2

This excellent primer covers solar heating principals, greenhouse design and construction, environmental control, cold frames and hotbeds and growing plants in the solar greenhouse. It's available through your local county agricultural extension agent.

Plants for Profit – A Complete Guide to Growing and Selling Greenhouse Crops, by Dr. Francis Jozwik.

This comprehensive guide is written for the first-time commercial grower. It explains in an easy-to-understand way how to successfully grow plants and flowers in the greenhouse, covering both what to grow and how to grow it. Special emphasis is placed on bedding plants, flowering pot plants and foliage plants. From : Andmar Press, P.O. Box 217, Mills, WY 82644. Dr. Jozwik also publishes a newsletter, "The National Greenhouse Grower".

Greenhouse Vegetable Production

Covers greenhouse location, structures, crop and cultural practices for commercial growers. $1

Greenhouse Tomato Production

Covers cultural practices and problems. $1

Greenhouse Cucumber Production
Covers flowering, fruit set, varieties, cultural practices, yields, harvesting, storage, pests and diseases. $2.50

These three handbooks are available from: Publications Office, Division of Agricultural and natural Resources, University of California, 6701 San Pablo Ave., Oakland, CA 94608-1239. Be sure to get a copy of their free publications catalog too!

If you're a greenhouse grower, or planning to be, the premier magazine in the field is **Grower Talks.** Monthly issues cover bedding plants, potted plants, foliage, cut flowers and vegetables.

New production techniques, plant marketing and new crops are examined every month in feature articles, and the annual directory is the "yellow pages" of the greenhouse industry. For subscription information, write: Grower Talks, Circulation Dept., P.O. Box 532, Geneva, IL 60134.

The Commercial Greenhouse Handbook by James W. Boodley.
This 568 page manual offers comprehensive advice of all aspects of greenhouse growing and management. Included is specific advice on flowering plants, bulbs, tropical plants, bedding plants and cut flower crops. Available through: AG-ACCESS, P.O. Box 2008, Davis, CA 95617

GREENHOUSE RESOURCE GUIDE
Greenhouse Equipment and Supplies —

Charley's Greenhouse Supply
1569 Memorial Hwy., Mt. Vernon, WA 98273
(Catalog $2)

Clover Garden Products
P.O. Box 789, Smyrna, TN 37167
Toll-free 1-800-251-1206

E.C. Geiger
Route 63, Box 285, Harleysville, PA 19438
Toll-free 1-800-443-4437

Hoop House
Fox Hill Farm, 20 Lawrence St., Rockville, CT 06066
Hoop House sells budget kits for a simple 10' wide greenhouse that can be built in any length from 8' to 48'. Catalog $1.

A.M. Leonard, Inc.
6665 Spiker Road, Piqua, OH 45356
Toll-free 1-800-543-8955

Mellingers
2310 S. Range Rd., North Lima, OH 44452, Toll-free 1-800-321-7444
The medium-sized hoop-style greenhouse kits offered here could be the perfect size for the beginning commercial grower. Stock widths are 12', 14', 18' and 25', lengths are expandable up to 148'.

Northern Greenhouse Sales, Box 42, Neche, ND 58265
Offers a unique 9 mil thick woven poly greenhouse covering that withstands hail and high winds and wears like iron. Send $1 for fabric sample and brochure.

Solar Components
P.O. Box 237, Manchester, NH 03105
Toll-free 1-800-258-3072
Offers components, kits and plans for solar greenhouses.

Stuppy Greenhouse Manufacturing
1212 Clay St., North Kansas City, MO 64116
Toll-free 1-800-821-2132

Greenhouse Organizations –

The Professional Plant Grower's Institute is a network of greenhouse growers and allied trades working for their mutual benefit. Membership has many advantages to both beginning and established growers, including: Training programs, marketing aids such as wall charts, full color plant labels, an inexpensive bedding plant information chart, customer education sheets, a monthly newsletter, reference books, tradeshows and conferences. Members also get an annual membership directory and buyer's guide. P.P.G.I., P.O. Box 27517, Lansing, MI 48909.

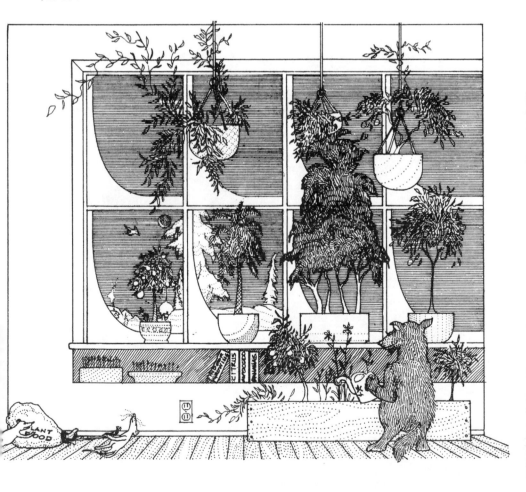

HYDROPONIC GROWING

Hydroponics is the practice of growing plants without soil. Plants are grown directly in a water/nutrient solution that contains the minerals necessary for plant nutrition and growth. At first glance, it might seem a bit strange to find plants thriving without soil. Yet growers all over the world are discovering that hydroponic growing has many advantages over conventional growing practices.

Hydroponic yields can be much greater than conventional growing yields or even greenhouse growing yields. One acre of hydroponic production has the potential growing capacity of 20-30 acres using conventional farming methods.

Since most hydroponic crops are grown in a greenhouse, the growing season is extended by several months, allowing multiple crops, and resulting in higher off-season prices to growers.

Inside the greenhouse environment, pests are much easier to control, virtually eliminating the need for costly insecticides. Most good growers use only an occasional biological control when necessary.

One benefit readily appreciated by anyone who has had a garden is that no weeding is required in hydroponic growing. In addition, the heavy labor of working the soil is eliminated. Most growers use a lightweight aggregate such as peat or perlite, which is practically weightless.

To get a better idea of what commercial hydroponics is all about, let's take a brief look at two successful growers. In the Blue Ridge mountains of North Carolina, Linda Caudill recently retired from nursing to grow hydroponic lettuce. She chose a package greenhouse and growing system from Clover Garden. She and her family are able to produce 3000 heads of lettuce a week, selling it to local grocery stores, restaurants, hospitals, and school cafeterias. Best of all, the demand for her hydroponic lettuce already exceeds the growing capacity so another greenhouse is being installed.

Their work schedule involves seeding, transplanting, and harvesting. The lettuce seeds are placed in special growing cells and misted for one week. Then the trays of seedlings are moved to a "nutrient-flow" table for another two weeks before transplanting. Then the individual plants are transplanted to large growing tables, similar to plastic gutters, where they continue to grow for another three or four weeks, depending on the season. At harvest time, the mature heads of lettuce are individually packed in a poly bag and boxed for delivery.

In Massachusetts, grower Pieter Schippers and his family grow a variety of hydroponic vegetables and flowers for market in three large 34' x 145' greenhouses. The kitchen garden is there too, after they discovered how much easier it was to grow hydroponically, with a large assortment of vegetables, herbs, and flowers for the table.

The plants are rooted in shallow beds of white perlite with a nutrient solution circulating through it. Tomatoes and cucumbers twine up strings tied to the overhead frame of the greenhouse.

According to the Schippers, the greenhouses cost about $15,000 each, or $3 a square foot. Gross yearly sales from each greenhouse are about $40,000 with a net profit after expenses of about $24,000 per greenhouse. This is almost $5 a square foot, equal to a very intensive outdoor growing system.

Each greenhouse is filled with waist-high racks, covered with nine foot sections of white plastic gutter. Each gutter is filled with perlite and contains plants spaced along the length of the gutter. At one end of the greenhouse the new seedlings are placed in gutters close together. Then as the plants grow, the gutters are pushed down the rack and spaced further apart to allow more room for growth. At the other end of the greenhouse, mature plants are harvested and packed.

Schippers claims this moveable gutter system can increase production by as much as 35 percent.

The growing method used by many commercial growers is called the Nutrient Flow Technique, or NFT. In the NFT method a nutrient solution flows continuously in the bottom of the plant growing channel. The solution is checked periodically, and nutrients are added as needed.

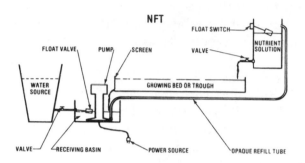

As you can see in the illustration above the nutrient solution flows from the upper tank through a sloped growing bed that is filled with perlite. The solution runs into a receiving tank at the far end of the growing bed. The system becomes automatic with the float switches in both tanks. One activates a recirculating pump to send the nutrient solution back to the upper tank. The second, connected to an outside water source such as needed to make up for evaporation and plant transpiration.

Bob Cheves, of Delray, West Virginia, got into hydroponics eight years ago when his seventh-grade daughter needed a science project! Since then, Bob has expanded to four 30'x100' greenhouses producing cucumbers and lettuce.

Bob raises European burpless cucumbers which grow from 12"-15" long and weigh just over a pound. One greenhouse produces 20,000 cukes yearly. The other greenhouse produces 40,000 to 50,000 heads of bibb lettuce each year.

Bob says that hydroponically grown lettuce and cucumbers have a different appearance from field-grown, so he labels each cuke with a special label and bags his lettuce to identify it as hydroponic. Once a shopper buys hydroponic produce, he can easily recognize it again. Bob uses a broker, Waterfield Farms, to market all his production, freeing him to concentrate on growing the best produce possible.

Most hydroponic growers agree that leaf crops offer the highest return for the commercial grower. Profitable leaf crops include leaf lettuce, bibb lettuce, endive and spinach. These crops are also easy to grow, requiring less effort and knowledge than tomatoes, cucumbers, and peppers. In addition, because of the harvest cycle, income is steady year-round.

According to the experts at Crop-King, "Lettuce is one of the major drawing cards for any good supermarket produce department. Being locally grown, your lettuce will only be a day or two from the greenhouse to salad bowl". This "hand-to-mouth" production results in a fresher product, a quality which has previously limited sales of leaf and bibb lettuce. It simply can not be bought fresh much of the year in stores.

Also, hydroponically grown lettuce and other leaf crops, including spinach, are perfectly clean, uniform, very attractive, and much larger than fieldgrown leaf crops. This makes them much more appealing to the consumer and the produce buyer.

The projected return for hydroponic lettuce, according to Crop King, from a typical 30' x 124' greenhouse is $57,000, with a profit after expenses of $43,000. The same size greenhouse, growing hydroponic tomatoes, has a projected return of $28,000 with a profit after expenses of $21,000.

HYDROPONICS RECOMMENDED READING

A new Canadian magazine is being published for the hobbyist and small commercial grower. Called the **21st Century Gardener**, it covers the field using a practical, down-to-earth approach, using well-illustrated articles to show the reader "how-to". A sample issue is $4 from: 21st Century Gardener, P.O. Box 189, Princeton, British Columbia, Canada VOX 1W0.

HYDROPONIC RESOURCES

The companies listed below supply a wide range of products and services to the commercial hydroponic growers. All offer complete packages for new growers, including greenhouses, equipment and supplies, training programs, and systems for small-scale growers.

Crop King, Inc.
P.O. Box 310, Medina, OH 44258

Clover Gardens
P.O. Box 789, Smyrna, TN 37167
Toll-free 1-800-251-1206

Hydro-Gardens, Inc.
P.O. Box 25058, Colorado Springs, CO 80936
Toll-free 1-800-634-6362

Hydroponic Growing Systems
32 Richardson Road, Ashby, MA 01431

Western Water Farm
1244 Seymour Street
Vancouver, British Columbia, Canada V6B 3N9

MARKETING YOUR CROPS

Growing a cash crop is only half the job for a grower. The second half is selling what you grow. For most growers this will mean direct marketing. Direct marketing is usually the simplest and least expensive way to sell your crops.

Before getting started, it is essential to think about a marketing plan. Contact your local county extension agent for assistance in developing a plan. Here are a few questions to ask which should give you a better idea of how to best market your particular crop.

1. How much land is available for growing?
2. How much money is available until harvest time?
3. How close is your land to a population center?
4. What is the existing demand for the crops you want to grow?
5. Do you enjoy dealing with people?
6. List all possible sales outlets for the crops you want to grow.

Once you have established the range of marketing possibilities, you can make a decision based on the facts about what types of marketing will work best for you. There are eight basic options for backyard cash crop marketing. Let's take a brief look.

FARMERS' MARKETS

According to a nationwide survey recently conducted by American Vegetable Grower magazine, farmers' markets are coming back fast in popularity. There are now more than 15,000 nationwide, growing at 15 percent a year, with sales at existing farmers' markets up over 40 percent per year!

The experts say the reason for this surge in popularity is customers who are looking for quality, variety, and freshness. They want to buy locally grown, fresh and seasonal produce, and are willing to pay well for it.

Susie Davis, a gardener in Dubois, Illinois, has been selling her produce at Farmers' Markets since 1977. Here's what she has to say about "tailgate marketing".

"I had planted a good-sized garden and felt I would have something extra to sell. Seeing an article in the local paper about a farmer's market forming in town only prompted me to plant more! The first Saturday I was ready with some cabbage, head lettuce and a batch of home-made cookies. Not sure if I could sell it all or bring it back home, I headed for town. The other vendors and I were sold out within an hour or so. As the summer progressed, I realized we weren't growing enough of a wide enough variety of vegetables.

The next spring, I planted more of everything I had grown the year before, the basics: tomatoes, peas, beans, cabbage, potatoes, etc. I also added new vegetables that had not been growing for my family, but planted just for market like okra, greens, sugar peas, and also rhubarb, asparagus, and small fruits for later years.

I have been able to sell most all I could grow on the land planted in small fruits and vegetables. If you are depending on small local markets to sell all you produce, you must pay close attention to customer demand so you don't overplant on items they won't buy in large quantities. Plant only what they will stand in line to buy.

As we all know, most home gardens are planted for main season vegetables and are not in production for a long season. The market gardener, if they planned ahead for the early and late season, and also for vegetables not easily grown in a small home garden, could sell a good amount of produce as soon as the public knew they were there. In small towns, the basic crops are still the most sought after either for fresh eating or, if price is right, for canning.

I have tried to plant as wide and diversified variety of vegetables and small fruits as possible. Not always in large quantities, but to plant enough in spaced plantings and cultivar choices to carry each item as long as the season will permit. This will help me to put as large as possible selection produce on my stand each week. By doing this, I'm almost certain to have at least one item each customer may want. If you can get a customer to buy one item from you, he will probably buy more. Better yet, if the customer is satisfied when he or she gets home, they will most likely stop at your stand each week to see what you have to sell."

If you are just getting started selling your crop or live off a well traveled road, a farmers' market gives you a chance to try "tailgate marketing" with little risk. Here are a few tips from the pros to help you succeed.

• Bring a folding table to display your produce. Be sure to leave a space at the edge of the table for customers to set packages while getting out their money!

• Bring enough small change and dollar bills to start the day. As you sell, put the larger bills in a safe place, such as in a money belt or hidden in your truck. Otherwise, a large stack of bills may attract the wrong kind of customer.

• Use a white or off-white poster board and felt-tip pens to make your display signs. Green, red or blue are better colors than black.

• If you don't have a certified scale, sell your produce by the piece or by volume.

• Price is not everything. More important is quality, freshness, flavor and clean, attractive produce.

• If your produce is suitable, offer samples for tasting. According to a recent survey, 70 percent of those who try will buy!

• Customers like to have posted prices to allow comparison shopping.

• It's easier to lower prices than raise them. Don't start to low. Two sure signs that your prices are too low are: no complaints, or you sell out too early.

• Use color contrast to improve your display. Put a row of green produce such as lettuce or cukes in between other items to perk up the "eye-appeal". Check the produce department at a large supermarket to see how the pros do it.

FOOD SUBSCRIBER NETWORK

Over the past decade, an Alabama scientist has developed a remarkable plan for small growers who want a guaranteed market for their crops. He has added a unique "twist" to the standard pick-your-own plan.

According to Dr. Booker T. Whatley: "The clientele membership club is the lifeblood of the whole setup. It enables the grower to plan production, anticipate demand, and of course — have a guaranteed market. The grower has to seek out people — city folks mostly — to be members of the club. The annual membership fee, $25 per household, gives each of the families the privilege of coming to the farm and harvesting produce at about 60 percent of the supermarket price."

Dr. Whatley also says: "We're bringing up children today who don't even know how vegetables or chickens are raised! So some parents see a farm visit as a wholesome and pleasant educational experience for their youngsters — one that the entire family can share. The average middle-class city person likes a chance to get out on a farm. It's a form of entertainment, and those folks can save money while they're having a good time."

A grower who is currently using the Whatley Plan, Frank Randle, says: "At first I couldn't believe that people would pay me money to harvest my crops, but I've found that they will."

Dr. Booker T. Whatley has put his years of experience into a new book "How To Make $100,000 Farming 25 Acres." The book tells you how to:

1. Build a guaranteed market for all your crops.

2. Gross a minimum of $3,000 per acre per year.

3. Enjoy year-round daily cash flow.

4. Have full-time year-round employment.

You can order this new Rodale Press book at any bookstore or check out a copy at your library.

Dr. Whatley and Tom Monaghan, owner of the Domino's Pizza chain and the Detroit Tigers, are now setting up a demonstration farm at Ann Arbor, Michigan. If you're planning to be in the area, stop by for a tour. Call (313) 995-4500 for directions.

For more information about this unique and increasingly popular marketing method, as well as subscription information for Dr. Whatley's "Small Farm Technical Newsletter", write to: Whatley Farms, P.O. Box 250027, Montgomery, AL 36125

MAIL-ORDER MARKETING

Mail-order can be an ideal marketing method if you have a suitable product. Possibilities include flower bulbs, seeds, herb plants, dried herbs, nuts (especially regional specialties like filberts), nursery stock, tree seedlings, dried mushrooms, bamboo, holly in season, and specialized products such as herbal sachets, jams and jellies.

Mail-order can be about as predictable as a trip to Las Vegas, so proceed with caution! Don't purchase expensive display ads in magazines, as they are far too risky and expensive for the beginner. The proven low-cost, low-risk way to start in mail order is to use classified ads that offer a free brochure about your products. As you build a customer list, remember that most first sales are break-even at best, and that your goal is to turn the first-time buyers into repeat customers who come back year after year. If you're serious about selling your crops or products by mail, get a copy of "Mail-Order Moonlighting", by Cecil Hoge, Sr. This paperback is by far the best book for the beginner, just packed with practical, how-to information.

PICK-YOUR-OWN

Marketing your crops on a U-pick basis benefits both you and your customers. You can eliminate the cost of harvesting, washing, grading, packing, refrigeration and transportation. Your customers get fresh-picked flavor and save money over the supermarket price.

All crops are not suitable for P.Y.O. harvest. Good P.Y.O. crops have high labor requirements, a relatively high value per acre, and may be quite perishable.

The best crops for P.Y.O. include strawberries, raspberries, blackberries, blueberries, grapes, apples, cherries, selected vegetables such as beans, peas, asparagus, greens and Christmas trees. Pumpkins, sweet corn, tomatoes and melons are popular market crops that are usually grower harvested and displayed for customers to "choose your own".

Another consideration for a suitable P.Y.O. crop is the time and money required for each crop. Vegetables, for example, can be harvested in one season, while strawberries require two years; grapes, three years; blueberries and apples, three to five years.

According to the marketing specialists at most agricultural colleges, P.Y.O. is likely to be the fastest growing form of direct grower-to-customer marketing in the next decade. A recent survey of established P.Y.O. growers has identified the major success factors for establishing a P.Y.O. operation. Keep in mind that not all these conditions must be met, but the more the better.

Success Factors for PYO Growers

1. Plan on long hours and hard work during the growing/harvesting season. Reward yourself with a two-month vacation this winter!

2. It's important to like people, as you'll be dealing with lots of them every day.

3. Using the family members to help out keeps overhead low — remember to share the profits with all of them.

4. If your P.Y.O. operation expands beyond five acres, you will need to hire employees. Above all else, find help that is trustworthy and dependable.

5. Give your customers personal, individual attention.

6. Be courteous and friendly.

7. When establishing your P.Y.O. operation, remember that most of your customers will live within a half-hour drive from your fields. Make sure the population base is there before you begin.

8. Whatever you grow, make sure it is the very best you can make it. Customers will repeat year after year if you grow top quality.

9. Neatness pays. A customer's first impression of your operation is lasting.

10. Provide adequate parking — most operators say ten cars is a minimum.

11. Be prepared for overflow crowds occasionally.

12. Advertise to let people know about your farm. You may grow the best crops in the county, but you must have customers to buy them. Surveys have shown that most customers, learn about your P.Y.O. operation from other customers, friends and neighbors. When you are getting started, however, you cannot rely on just word-of-mouth advertising.

Classified advertising is the next best approach for reaching a large audience for the least amount of money. Use weeklies in smaller towns, "shoppers" in large communities. Other good advertising possibilities include free feature stories in local papers, radio spots, postcards to last year's customers (use a sign-in book at your checkout area to develop a mailing list), road signs, and advertising on your bags and containers.

In Sonoma County, California, the local growers' association got together to produce a map that makes it easy for the customer to visit and buy from the area's direct marketing farmers.

Their map, complete with all the back roads and participating farms marked on the map by number, includes two features that make it special. First, it includes a product directory referenced to farms; second, it includes a fresh produce calendar so customers can easily figure out what's in season and where to find it.

13. Become known for a specialty crop, such as pumpkins for Halloween.

14. Have attractive signs along the roads leading to your P.Y.O. field whenever possible, and then at the check-in/check-out station with prices, rules and any other necessary information.

15. A brochure is helpful, especially if you have many P.Y.O. crops — it could contain a picking calendar or recipes.

16. Keep regular hours and let people know about them.

17. Show your customers the best way to harvest and handle the crops.

18. Make your location easy to find.

19. Promote your operation for free publicity. For example, P.Y.O. apple orchards can have an apple butter festival in the fall, and an apple blossom walk in the spring. How about a Halloween pumpkin harvest for local school childrens' groups?

As P.Y.O. marketing has grown in the last few years, state extension specialists have devoted more time and attention to developing marketing plans for local crops and producers. Write to the publications office of your state agricultural university, listed in the next chapter, for a list of current publications on P.Y.O.

SUCCESSFUL HIGHWAY SIGNS

Your roadside sign must be readable to bring in the customers. According to a recent USDA study, 75 percent of the customers learned about a roadside stand from a highway sign. Remember that passing motorists have only a few seconds to see your sign and make a decision to stop. Here's how to make a readable sign:

1. Keep the message short. Six words is all most people can read while passing a sign.

2. Symbols and pictures are easier to understand than words. A single good picture or illustration of your crop is worth a thousand words.

3. Letters should be large enough so drivers can read your message. The "headline" should use letters 8-12" high. The "subheadline" should be in letters 4-6" high.

4. Use contrasting colors (colors opposite each other on the color wheel) for best visibility, but remember that red pigments fade rapidly in direct sunlight.

5. Make sure your sign is safe and legal. It should not block visibility, and it should meet local zoning ordinances for set-back and sign size.

6. Use an advance sign ¼ to ½ mile from your main sign to give drivers time to think about stopping.

RENT-A-TREE

The rent-a-tree concept is becoming quite popular across the country. Here's how it works. A grower with an orchard of fruit or nut trees wants to sell his crop before the harvest. A customer also wants to own a future crop while having a professional raise it.

Rent-a-tree can be profitable if you understand the potential of your crop and charge accordingly. It is also an opportunity to educate the local community by allowing folks to become involved in farming or orcharding.

To insure that everyone's expectations are met, it is customary for both grower and customer to sign a contract. A guide to preparing a rent-a-tree contract is available free from: Extension Marketing Specialist, Horticulture Department, Oregon State University Extension Service, Corvallis, OR 97331

ROADSIDE STAND

A seasonal roadside stand, located right on your farm, is an economical way to sell your crops. Many customers appreciate freshness, but just don't have the time or inclination to pick their own.

Your roadside stand can be as simple as you want, or as elaborate. Many growers have started with just an old card table and gone on to a year-round roadside market complete with walk-in coolers! A simple shed-type stand, clean and tidy, will encourage passing motorists to stop and shop. Several excellent do-it-yourself plans for affordable roadside stands are available through your local county extension agent.

A survey of roadside stand customers found that quality and freshness were the main reason for buying at the stands. Price rated below convenience and friendliness! The importance of quality, freshness, and friendliness is clear when you learn that most of the customers at roadside stands are repeat customers, shopping at the stands an average of once a week.

The Extension Service in each state usually holds an annual roadside marketing conference to provide the latest information on production and marketing techniques. Individual marketers are selected to describe their successful operation. Most conferences include demonstration booths for suppliers of equipment and goods, such as bags and baskets. Check with your local extension agent to find out when the next conference is scheduled in your state.

WHOLESALE OUTLETS

For those growers who prefer to spend more time growing, and less time selling, or whose specialty crop is simply too large or unique for direct marketing, wholesaling is the best marketing approach.

In addition to the usual brokers and wholesalers, consider institutional sales. For example, many restaurants, hospitals, schools, nursing homes, and hotels will buy direct from local growers whenever possible if the price and quality is right.

Growers of organic crops should get a copy of the annual directory "Wholesalers of Organic Produce and Products" published by the non-profit California Agrarian Action Project at P.O. Box 464, Davis, CA 95617. The directory includes state-by-state listings of growers, wholesalers, and buyer's co-operatives.

DRYING YOUR SURPLUS CROP

As a grower, no matter how well you plan, there will be part of a crop you just can't sell fresh. Perhaps the weather was perfect, leading to a bumper crop and a surplus. Perhaps every other market grower decided to plant a few extra tomato plants. Perhaps a windstorm gave you too many bruised windfalls in your apple orchard. Don't despair — Margaret Gubin can help you turn that "unsaleable" surplus into profits.

Margaret Gubin, a fifth generation farmer in Wisconsin, teaches food dehydration on 14 college campuses and on radio. She recently set up a plant called "Cherry Delite", which dries the surplus cherries in Northern Wisconsin.

Says Margaret, "What does the smart farmer put away for safekeeping in case of hard times? If you have wealth, you put away money. If you don't, you put away food. And the best food to put away is dried food, taking very little space and no refrigerator. All of this at a quarter of the cost of freezing and half the cost of canning.

The day has come when food drying is for more than self-defense against bad times. Drying can be used to create good times by adding income to the family. Fruits, vegetables, flowers, herbs, meats, and a large variety of crafts can be dried. Any grown commodity can be dried and resold for a profit. New markets are opening daily, and are limited only by one's imagination.

What if often wasted in other processes is often saved in drying. A bruise here, or an off-color may be wasted otherwise, but can be utilized in drying. Everything can be sold. I see no end to the possibilities for folks to add to their incomes with food drying technology, a little imagination, and a little hard work without leaving home."

For more information on food drying and a brochure on the "Equi-Flow" home and commercial food dehydrators, send a large stamped envelope to: Margaret Gubin, Margaret's Country Cupboard, N7160 County I, Juneau, WI 53039.

WHERE TO GET HELP

One of the little-known benefits of starting a growing business is the variety of free professional assistance available from government and university sources.

For instance, in many states the Extension Service, a branch of the U.S. Department of Agriculture, has established special programs geared towards helping small growers.

At the state level, many universities and colleges provide both technical and management assistance. Here is a complete state-by-state listing showing where to call or write for **free** expert plant advice in your area.

Alabama – Auburn University, Alabama Cooperative Extension Service, 116 Extension Hall, Auburn, AL 36849, (205) 826-4985.

Alaska – Wayne Vandre, 2221 E. Northern Lights Blvd., No. 240, Anchorage, AK 99508, (907) 279-5582.

Arizona – University of Arizona, Department of Plant Sciences, Tucson, AZ 85721, (602) 621-1945.

Arkansas – University of Arkansas, Department of Horticulture and Forestry, Plant Science 314, Fayetteville, AR 72701, (501) 521-2603.

California – University of California, Davis, Department of Horticulture, Davis, CA 95616, (916) 752-6617.

Colorado – Colorado State University, Department of Horticulture, Fort Collins, CO 80523, (303) 491-7119.

Connecticut – University of Connecticut, Department of Plant Science, 1376 Storrs Road, Storrs, CT 06268, (203) 486-3435.

Delaware – University of Delaware, Department of Horticulture, Townsend Hall, Newark, DE 19717-1303, (302) 451-2506.

Florida – University of Florida, Department of Ornamental Horticulture, IFAS, 1545 Fifield, FL 32611, (904) 392-7935.

Georgia – University of Georgia, Department of Horticulture, Athens, GA 30602, (404) 542-2340.

Hawaii – University of Hawaii, Department of Horticulture, Manoa, 3190 Maile Way, Honolulu, HI 96822, (808) 948-8909.

Idaho – University of Idaho, Department of PSES, Moscow, ID 83843, (208) 885-6276.

Illinois – University of Illinois, Department of Horticulture, 1013 Plant Sciences Lab, 1201 S. Dorner, Urbana, IL 61801, (217) 333-2124.

Indiana – Purdue University, Department of Horticulture, West Lafayette, IN 47907, (317) 494-1335.

Iowa – Iowa State University, Department of Horticulture, 105 Horticulture Bldg., Ames, IA 50010, (515) 294-0029.

Kansas – Kansas State University, Umberger Hall, Manhattan, KS 66506, (913) 532-5820.

Kentucky – University of Kentucky, Department of Horticulture and Architectural Landscaping, N318, Agricultural Science Center N., Lexington, KY 40546, (606) 257-4721.

Louisiana – Louisiana State University, Department of Horticulture, Knapp Hall, Baton Rouge, LA 70803, (504) 388-2222.

Maine – University of Maine, Department of Horticulture, 119 Deering Hall, Orono, ME 04469, (207) 581-2949.

Maryland – University of Maryland, Department of Horticulture, 2107 Holzapfel Hall, University Park, MD 20742, (301) 454-8924.

Massachusetts University of Massachusetts, Department of Plant and Soil Science, French Hall, Amherst, MA 01003, (413) 545-0895.

Michigan – Michigan State University, Horticultural Department, East Lansing, MI 48824, (517) 355-5178.

Minnesota – University of Minnesota, Department of Horticultural Sciences and Landscape Architecture, St. Paul, MN 55108, (612) 624-9703.

Mississippi – Mississippi State University, Department of Horticulture, P.O. Drawer T, Mississipi State, MS 39762, (601) 325-3223.

Missouri – University of Missouri-Columbia, Department of Horticulture, 1-40 Agriculture Building, Columbia, MO 65211, (314) 882-7511.

Montana – Cooperative Extension Service, Taylor Hall, Bozeman, MT 59717, (406) 994-3402.

Nebraska – University of Nebraska, Department of Horticulture, 377L Plant Science Building, Lincoln, NE 68538, (402) 472-1145.

Nevada – University of Nevada, Department of Plant Science, Reno, NV 89557-0107, (702) 784-6911.

New Hampshire – University of New Hampshire, Department of Horticulture, Nesmith Hall, Durham, NH 03824, (603) 862-3207.

New Jersey – Rutgers University, Department of Horticulture and Forestry, P.O. Box 231, Blake Hall, Cook College, New Brunswick, NJ 08903, (201) 932-8424.

New Mexico – New Mexico State University, 9301 Indian School Rd. NE, Suite 101, Alburquerque, NM 87112, (505) 275-5231.

New York – Cornell University, Department of Horticulture, Ithaca, NY 14853, (607) 255-2166.

North Carolina – North Carolina State University, Department of Horticultural Science, Kilgore Hall, Raleigh, NC 27695, (919) 737-3322.

North Dakota – North Dakota State University, Department of Horticulture, Box 5658, University Station, Fargo, ND 58105, (701) 237-8161.

Ohio – Ohio State University, Department of Horticulture, 241 Howlett Hall, 2001 Fyffe Court, Columbus, OH 43210, (614) 292-9784.

Oklahoma – Oklahoma State University, Department of Horticulture, Stillwater, OK 74078, (405) 624-5414.

Oregon – Oregon State University, Department of Horticulture, Corvallis, OR 97331, (503) 754-3464.

Pennsylvania – Pennsylvania State University, Department of Horticulture, 102 Tyson Bldg., University Park, PA 16802, (814) 865-6596.

Puerto Rico – University of Puerto Rico, Department of Horticulture, Mayaque Campus, Puerto Rico 00708, (809) 832-4040.

Rhode Island – University of Rhode Island, Department of Plant Sciences, Kingston, RI 02881, (401) 792-2791.

South Carolina – Clemson University, Department of Horticulture, Clemson, SC 29634, (808) 656-2603.

South Dakota – South Dakota State University, Department of Horticulture, Brookings, SD 57007, (605) 688-5136.

Tennessee – University of Tennessee, Department of Ornamental Horticulture, Box 1071, Knoxville, TN 37901, (615) 974-7324.

Texas – Texas A & M University, Department of Horticulture, College Station, TX 77843, (409) 845-5341.

Utah – Utah State University, Department of Horticulture, Logan, UT 84322-4820, (801) 750-2258.

Vermont – University of Vermont, Plant and Soil Science Department, Hills Bldg., Burlington, VT 05405, (802) 656-2630.

Virginia – Virginia Polytechnic Intitute and State University, Department of Horticulture, 301 Saunders Hall, Blacksburg, VA 24061, (703) 961-5451.
Bonnie Appleton, Virginia Tech., Extension Division, Hampton Roads, Agriculture Experimental Station, 1444 Diamond Springs Rd., Virginia Beach, VA 23455, (804) 446-4906.

Washington – Washington State University, Department of Horticulture, Pullman, WA 99164-6414, (509) 335-9502.

West Virginia – Cooperative Extension Service, West Virginia University, P.O. Box 6031, 817 Knapp Hall, Morgantown, WV 26506-6031, (304) 293-5691.

Wisconsin – University of Wisconsin, Department of Horticulture, 1575 Linden Dr., Madison, WI 53706, (608) 262-1450.

Wyoming – University of Wyoming, Box 3354, University Station, Laramie, WY 82071, (307) 766-2243.

At the federal level, a wide variety of services, such as technical assistance, marketing advice, loans and loan guarantees, and information is available to you at no charge.

The U.S. Government is the most comprehensive, yet most unexplored and under-used source of information in the world. Of the several million federal employees, nearly 30 percent are information specialists, just waiting for your call or letter!

As a taxpayer, you pay their salaries and fund their research. As a citizen-grower, you are entitled to a return on your tax investment. Here are several programs and services of the federal government that may help you as a grower/small businessperson.

• The U.S. Department of Agriculture lends up to $30,000 to rural youths from 10-21 years of age. The loans must be used to start income-producing projects such as backyard cash crops or a roadside stand. Contact: Production Loan Division, Farmers Home Administration, USDA, Washington, DC 20250

• Free technical assistance is available to help you diagnose and cure plant diseases. Services range from telephone consultation to leaf and soil analysis. Contact: your local extension service office.

• The USDA has staff experts that can supply marketing information on almost any crop you could imagine. For example, if you're thinking of growing oyster mushrooms and need to know how large the market is, how fast it's growing, and where to sell your crop, they can help. Contact: Marketing Research and Development Division, USDA, AMS, Room 130, Bldg. 307-BARC-E, Beltsville, MD 20705

• A large USDA research staff is involved in areas such as crop production, plant genetics, plant nutrition, and farm products. For free technical expertise on your specific interest, contact: Agricultural Research Service, USDA, Room 302-A, Administration Bldg., Washington, DC 20250

• A wide variety of assistance is available to growers who want to export their crop or agricultural products. This ranges from a weekly bulletin, "Export Briefs", listing crops wanted by foreign purchasers to export marketing research and financing of export sales for US growers. Contact: Foreign Agricultural Service, USDA, 14th Street & Independence Ave. S.W., Washington, DC 20250

• The USDA crop insurance program is available to growers nationwide, and covers drought, freezes, insects and other natural causes. Contact: Manager, Federal Crop Insurance Corporation, USDA, Washington, DC 20250

• The USDA has two loan programs available for growers. The first, farm operating loans, provides credit for growers who are unable to obtain credit elsewhere. The loans are up to a maximum of $200,000. Contact: Director, Production Loan Division, Farmers Home Administration, USDA, Washington, DC 20250
The second program, farm ownership loans, operates on the same basic principle, but provides purchase loans for land or a barn or greenhouse. Contact the office listed above.

• The National Agricultural Library at the USDA has a team of researchers available at no cost to answer specific questions or provide information on almost any agricultural topic. Contact: National Agricultural Library, 10301 Baltimore Blvd., Beltsville, MD 20705

• In addition, a USDA staff of research specialists is available to get you specific answers to your questions or steer you to an expert who can help. Even if the USDA does not have the answer, they can help you locate someone who will. Contact: Information Office, Office of Public Affairs, USDA, Room 230E, Washington, DC 20250

• The U.S. Small Business Administration provides assistance and loans to small businesses, including growers. Their loans can be used for working capital, to purchase inventory, equipment and supplies, or for construction and expansion. They offer two types of business loans: Loans made by private lenders, such as banks, and guaranteed by the SBA, and loans made directly by the SBA. A free booklet, "Business Loans from the SBA", is available with up-to-date information about their loan programs. Contact: Office of Business Loans, SBA, 1441 L Street NW, Room 804D, Washington, DC 20416

• The SBA has hundreds of free publications available, covering such topics as small business management, marketing and financing a business. Contact: Publications Department, SBA, 1441 L Street NW, Room 100, Washington, DC 20416

• S.C.O.R.E., the Service Corps of Retired Executives, volunteer their services to help small businesses solve their problems, or start a new business. For more information about this excellent program, contact: SCORE, Management Assistance, SBA, 1441 L Street, NW, Room 602H, Washington, DC 20416.

• Small Business Development Centers. Individual counseling and training are available at no cost to small business owners. Talent is drawn from government resources, university facilities, and private resources to provide technical and managerial assistance. Contact: Small Business Development Center, management Assistance, SBA, 1441 L Street NW, Room 602, Washington, DC 20416.

• Small Business Institute. Extensive business consulting if offered, free of charge, through over 500 universities and colleges, to small business owners. If you want to expand the market for your crops or learn how to do a better job of running your business, contact: Small Business Institute, management assistance, SBA, 1441 L Street NW, Room 602H, Washington, DC 20416.

RECOMMENDED READING

Gardening by Mail, by Barbara Barton, is a huge directory of "everything for the garden and gardener." It contains mail-order sources for gardeners in both the U.S. and Canada, including over 1000 seed companies, 360 garden supply and service companies, 230 plant and horticultural societies, and over 200 gardening and horticultural magazines and newsletters. What a book! Get a copy at your local bookstore or library.

In addition to the suppliers listed in previous chapters for seeds and plants, rare and hard-to-find plants or seeds can often be found through horticultural societies, such as the ones listed in "Gardening by Mail". For example, the American Bamboo

Society has an annual sale of exotic bamboo each year at the Quail Botanic Gardens. The members of most groups are very generous with seeds and cuttings.

Small-Scale Agriculture Today is the title of a new newsletter put out by the USDA office of small scale agriculture. It covers the spectrum of small-scale growing possibilities, news of seminars, workshops, new books and new products and crops. Send for a free sample issue. USDA/CSRS/SPPS Office of Small Scale Agriculture, 14th and Independence SW, Washington, DC 20251-2200.

If you're having trouble finding any books listed in previous chapters, or any specialized book about gardening, agriculture or horticulture, here are mail-order sources that specialize in those areas:

Ag-Access, 603 4th St., Davis, CA 95617 (919) 756-7177

I.S.B.S., 5602 NE Hassalo St., Portland, OR 97213, (503) 287-3093, Toll-free (800) 547-7734

American Nurseryman, 111 N. Canal St., No. 545, Chicago, IL 60606, Toll-free (800) 621-5727

Rural Enterprise Magazine covers the latest developments in small farm diversification, direct marketing, and rural homebased businesses. Each quarterly issue provides feature stories about successful pick-your-own operations, farmer's markets, specialty crops and livestock. A sample issue is $3. Rural Enterprise, P.O. Box 878, Menomonee Falls, WI 53051.

New Farm Magazine was founded by the "grandfather" of the organic gardening and agriculture movement, Robert Rodale, and points the way to a profitable farm future that is sustainable and chemical-free. Aimed at farmers with more than an acre or two, it features new ways to diversify for profits. Sample issue $3, from; New Farm, 222 Main Street, Emmaus, PA 18098.

Because organic or chemical-free growing is rapidly becoming the preferred way to grow, the need for organic fertilizers and biological and natural pest controls has increased as well. Two excellent mail-order sources are:

Peaceful Valley Farm Supply, 11173 Peaceful Valley Rd., Nevada City, CA 95959 (916) 265-3276. Catalog $2.

Necessary Trading Co., 683 Main Street, New Castle, VA 24127 (703) 864-5103. Free Catalog.

Because so many growers are switching to chemical-free growing techniques, the U.S. Department of Agriculture has established an "Organic Agriculture Hot Line". The information center has set up a toll-free access number (800) 346-9140, and is staffed by experts in soil sciences and pest control. The lines are open 8 a.m. to 5 p.m. Central Time.

RUSH ORDER FORM

RUSH ORDER FORM

PRINT YOUR NAME _____

YOUR ADDRESS _____

CITY _____

STATE _____ ZIP _____

   *Special Reports–please order by number.
   Each report is $9, any three for $21

   Reports ____, ____, ____, ____, ____, $_____

   ____, ____, ____, ____, ____, ____,

   *Small Barn Planbook $3 $_____

   TOTAL YOUR ORDER HERE
      (all prices include postage) $_____
   Do you live in Washington State?
   Yes ____ Lucky you! Add 8% state sales tax $_____

SHIPPING. We make every effort to fill your order promptly.
Special reports are sent out by First Class mail.

Mail your order to: **Specialty Crop Digest**
                       Post Office Box 1058
                       Bellingham, WA 98227

SPECIALTY CROP DIGEST

The Specialty Crop Digest is jam-packed with money-making and money saving news and ideas for market growers, big and small. Write today for your sample copy of this digest. Be sure to include a dollar for first class postage and handling. Mail to: P.O. Box 1058, Bellingham, WA 98227.

SPECIAL REPORTS

Our popular special reports on high-value crops are written to give you all the basic "start-up" information to plant, grow, harvest–and make money. Each report is 10-15 pages of solid information you can use now. Don't expect fancy packaging–just hard-to-find, down-to-earth information to help you prosper as a grower. All special reports are just $9 each, or stock up at 3 for $21.

THE SWEET SMELL OF SUCCESS

How to make $2 per square foot growing garlic.

"There is a booming market out there for local, fresh garlic. Those growing it now sell every clove they can produce. Elephant garlic retails for $6 a pound, and produces at the rate of 15,000 pounds per acre."

Roger Kline
Vegetable Crops Specialist
Cornell University

Garlic is an ideal crop for the small grower, as it's labor-intensive and almost foolproof. Because garlic tolerates a wide variety of soils and weather, it is very, very hard to lose a crop. For decades, growers have called it "the mortgage lifter" for that very reason. This report covers all the essential information you'll need - proper soil preparation for intensive plantings, hard-to-find sources for exotic garlic seed cloves, planting for a bumper crop, how to keep your crop healthy. When and how to harvest for premium bulbs, how to double the prices most growers get for their bulbs, sources for mesh bags, the ten best markets for your garlic and value-added garlic-based products. ORDER #R1.

CHRISTMAS TREES

How to make up to $40,000 an acre growing and selling trees, wreaths and greens.

Over 15,000 growers can't be wrong! According to the National Christmas Tree Association, most are part-timers, growing on one to twenty acres. In this report, you'll learn how to pick the best growing stock—including six wholesale mail-order sources for fast-growing seedlings. Planting for low maintenance and top dollar returns, how to feed to produce premium trees, when and how to shear and how to produce "living" container trees. Some growers make more money from wreaths, sprays, garlands and centerpieces than they do from Christmas trees—you'll learn how to do the same. The beauty of these "sidelines" is that you can start now—no need to wait for your trees to grow. There's a list of wholesalers for wreath-making supplies, and tips on selling your trees and greens with free advertising. One little-known method listed in the report shows you how to harvest and sell hundreds of trees without lifting a finger! ORDER #R2.

MAKING MONEY WITH OYSTER MUSHROOMS

Oyster mushrooms are an ideal cash crop for the grower with a limited budget and limited space. In just a 10 foot by ten foot space—a corner of the basement, garage or backyard shed—you can grow as much as 300 pounds a month of these exotic and tasty mushrooms. Best of all, retail prices are currently between $5 and $12 per pound! According to the U.S. Department of Agriculture, demand is growing for oyster mushrooms, with 1.5 million pounds sold last year. Unlike regular mushrooms, oyster mushrooms don't require the messy (and smelly) manure compost. Plain straw (wheat or rice straw work just fine) provides all the food they need. They're efficient eaters, too, converting a pound of straw into almost one pound of mushrooms!

This "gets you started" report will show you how to set up a growing area on a budget, where to buy spawn (seed), how to grow and harvest, and where and how to sell your harvest for top dollar. ORDER #R3.

ORNAMENTAL GRASS NURSERY

The ornamental grasses are enjoying a surge in popularity today. They range in size from low growing 6-inch tufts to 20-foot giants. Landscape designers love them because they can be used in so many ways–ground covers, specimen plants, in borders or near ponds and streams, as privacy screens and in rock gardens. Several cities are using ornamental grasses for urban landscaping because they are so tolerant of air pollution and poor soil.

Thanks to this growing popularity, nurseries are struggling to keep up with the demand. In this report, you'll learn which plants are "hot," and how to set up a profitable nursery in as little as a quarter-acre, how to tap the lucrative landscaping market and where to find wholesale seed and starts. Ornamental grasses are ideal for the beginning grower, as few insects or diseases bother them, and you can sell your first crop after just one season. ORDER #R4.

PROFITABLE CULINARY HERBS

How to make $1,000 a month part-time growing culinary herbs.

Not so long ago, it was impossible to find fresh herbs at the grocers or supermarket. Now, thanks to cooking magazines, TV cooking shows and a growing level of cooking sophistication, customers are asking for fresh herbs at the produce section.

This growing demand has created an opportunity for backyard herb growers. This report will show you exactly how to tap this demand and profit. The two most lucrative (and dependable) customers for the small grower. The ten best culinary herbs to grow–how to grow and harvest them. How to get started for less than $300. And for those who don't even have a garden–a special section on hydroponic herb growing in a spare room, basement or garage. One grower using this hydroponic setup is getting $5 per square foot of growing space–per month! His setup takes 240 square feet of basement and nets $1,200 per month. ORDER #R5.

COLORFUL PROFITS WITH DRIED FLOWERS

The fastest growing segment of agriculture today is flower production, says Dave Koranski, USDA horticulture specialist. Koranski says the demand is increasing each year by 10% or more. Dried flowers or everlastings are an ideal crop for backyard growers. As a group, they are easy to grow and easy to dry. Best of all, once you've dried your crop, there's no hurry to sell to avoid spoilage, as with fresh flowers. Your dried flowers will keep just fine until sold.

You'll learn the 15 most popular (and profitable) flowers to grow and how to grow, harvest and dry each one. You'll learn the six best markets for dried flowers—both wholesale and retail. You'll learn which colors are in demand—value added products you can make to double your profits—wholesale sources for seed and transplants.

If you're starting on a shoestring, consider growing everlastings. One grower I talked to recently was making $8 per square foot of growing area (wholesale) to $16 per square foot (retail). That's around $10,000 per season for an average 600 square foot garden. ORDER #R6.

SMALL BARN PLANBOOK

Growers everywhere can always use extra space—for storing equipment, drying the crop, starting a retail shop for more sales and so many other uses. The Homestead Design Planbook can help you provide that space—with a complete collection of 27 new designs, including gardens sheds, an expandable greenhouse, small barns, stables and garages.

Sizes range from 100 square feet to over 1,600 square feet. There are old-fashioned styles, like our gambrel and saltbox barns and garages, and the always-popular Monitor style western barns. Need a garage? Take your pick of fourteen different designs, some with full second floors. All buildings are designed with the owner-builder in mind so you can save by building it yourself. Even if you decide to have someone build it for you, these practical designs allow you to keep costs low. The planbook is just $3. ORDER #PB.